Managing Nuclear Accidents

Managing Nuclear Accidents

A Model Emergency Response Plan
for Power Plants and Communities

Dominic Golding, Jeanne X. Kasperson,
Roger E. Kasperson, Robert Goble,
John E. Seley, Gordon Thompson,
and Charles P. Wolf

Routledge
Taylor & Francis Group

NEW YORK AND LONDON

First published in 1992 by Westview Press

Published in 2021 by Routledge
605 Third Avenue, New York, NY 10017
2 Park Square, Milton Park, Abingdon, Oxon OX14 4RN

Routledge is an imprint of the Taylor & Francis Group, an informa business

Copyright © 1992 by Taylor & Francis

A CIP catalog record for this book is available from the Library of Congress.
ISBN 0-8133-8520-2

ISBN 13: 978-0-3670-0796-6 (hbk)
ISBN 13: 978-0-3671-5783-8 (pbk)

CONTENTS

LONG-TERM PROTECTIVE ACTIONS 103

MAINTAINING EFFECTIVE PREPAREDNESS 109

TABLES AND FIGURES

ABOUT THE AUTHORS

Dominic Golding is a fellow in the Center for Risk Management at Resources for the Future. He received his Ph.D in geography from Clark University where his research focused on occupational hazards and the social issues of risk assessment and risk management, especially with regard to nuclear power. His current research interests include the history and development of risk research, environmental equity, risk communication, and the evaluation of risk burdens in individual communities. Golding is also the author of *The Differential Susceptibility of Workers to Occupational Hazards: A Comparison of Policies in Sweden, Britain, and the United States* (1989) and contributing coeditor of *Social Theories of Risk* (1992).

Jeanne X. Kasperson is Research Librarian and Director of Publications at the George Perkins Marsh Institute at Clark University and senior Research Associate at the Alan Shawn Feinstein World Hunger Program at Brown University. Her recent research includes work on the social amplification of risk, risk communication, risk signals in the media, corporate culture, world hunger, and global environmental change. She currently serves on the editorial boards of *Environment* and *Risk Abstracts* and is book review editor for the latter. She is a contributing coeditor of *Water Re-Use and the Cities* (1977), *Risk in the Technological Society* (1982), *Perilous Progress* (1985), *Nuclear Risk in Comparative Perspective* (1987), and *Corporate Management of Health and Safety Hazards* (1988).

Roger E. Kasperson, who holds his Ph.D. from The University of Chicago, is coauthor or contributing coeditor of *Water Re-Use and the Cities*, (1977), *Equity Issues in Radioactive Waste Management* (1985), *Nuclear Risk Analysis in Comparative Perspective* (1987), *Corporate Management of Health and Safety Hazards* (1988), *Risk Communication: Proceedings of the International Workshop on Risk Communication, October 17-21, 1988* (1989), and *Communicating Risks to the Public* (1991). He has written widely on issues connected with technological hazards, risk communication, radioactive wastes, and global environmental change. His current research projects deal with the risk and social impacts associated with the siting of hazardous waste facilities, evaluation of risk-communication programs, global environmental change, and critical environmental zones.

Robert Goble is a Research Professor in Environment, Technology and Society at Clark University and a senior researcher at the Center for Technology, Environment, and Development (CENTED). He holds a Ph.D. in elementary particle physics, and for the past fifteen years he has conducted research and taught in the general areas of energy systems and policy, air quality (including indoor air quality), nuclear safety, and risk assessment. He has written and testifies extensively on nuclear economics, accident consequences, radioactive waste disposal, and nuclear accident emergency-planning efforts in the United States and Canada. His present research is on risk-assessment methodologies and on responses to the threat of global warming.

John E. Seley , Professor of Urban Studies at Queens College, City University of New York (CUNY) and Research Associate, Woodrow Wilson School, Princeton University, received his Ph.D. in city and regional planning from the University of Pennsylvania and has taught at the University of Minnesota and at Clark and Cornell Universities. At Queens he is also the founder and director of the Office of Community Studies. He has published widely on the subjects of equity in public service delivery, nuclear waste, and emergency planning. Seley is coauthor of a study of the feasibility of an independent Staten Island and author of *The Politics of Public-Facility Planning* (1983). His recent research centers on several national issues involving hazardous waste, community right-to-know, and emergency preparedness.

Gordon Thompson is Executive Director of the Institute for Resource and Security Studies, Cambridge, Massachusetts. Educated in science and engineering in his native Australia, he obtained a doctorate in applied mathematics from Oxford University in 1973. Since then he has pursued a wide-ranging career, performing technical and policy analysis on energy, environment, sustainable development, and international security. He has investigated the safety of nuclear facilities in the United States, the United Kingdom, Canada, and Germany. Since 1979 he has been based in the United States.

Charles P. Wolf is Director of the Social Impact Assessment Center in New York City and Adjunct Professor in the College of Environmental Science and Forestry, State University of New York, Syracuse. He was research director of the Social Science Research Council's study for the President's Commission on the Accident at Three Mile Island (the Kemeny Commission) and coeditor of the study report, *Accident at Three Mile Island: The Human Dimensions* (1982). Recently he has been involved in assessing the socioeconomic impacts of high-level nuclear waste and outer-continental-shelf oil and gas.

ACKNOWLEDGMENTS

This volume has benefitted from the financial support of the Three Mile Island Public Health Fund and its scientific advisory board, which provided comments and review throughout the course of the project. The authors are particularly indebted to Dr. Jonathan Berger, Executive Secretary of the Fund, for his sustained encouragement and intellectual nurturing of the research. We owe yet another salute to Lu Ann Renzoni-Pacenka, whose indefatigable perfection has turned out yet another camera-ready volume, all the while accommodating editorial license that ranged from the whimsical to the perverse. Somehow, her good humor is intact even though she knows full well that the second volume of this study is lurking in the wings and awaiting her expert touch.

Dominic Golding
Jeanne X. Kasperson
Roger E. Kasperson
Robert Goble
John E. Seley
Gordon Thompson
Charles P. Wolf

INTRODUCTION

An accident at a nuclear power plant can pose a serious threat to surrounding populations. If people and their governments are poorly prepared, as was the case at Windscale, England in 1957, Three Mile Island (TMI) in the United States in 1979, and Chernobyl in the Soviet Union in 1986, substantial exposure to radiation could occur. If, on the other hand, careful preparations have been made, timely and effective responses may reduce public exposure to radiation. In this sense, emergency planning can reduce, although not eliminate, the risks associated with nuclear power plants.

Much has been done since the 1979 TMI accident to prepare for a future accident in the United States. The federal government now requires approved emergency plans for all nuclear plants and has established minimum specifications for such plans. These plans, which delineate who will respond to an emergency at a nuclear plant and what their responsibilities will be, include objectives for public protection. Investments in communications systems aim to ensure that the public will be alerted in a timely way. Drills and exercises have been mounted to train those who will participate in the emergency response and to assess whether the plan will actually work.

For all this, major questions remain regarding the effectiveness of these plans in an actual nuclear accident, although they have performed well in non-nuclear incidents. Because serious nuclear accidents occur only very infrequently, the plans have not been tested in real circumstances. Also, since 1980, when the present federal regulations were published, much has been learned about nuclear accidents and the behavior of people during emergencies. Accordingly, the Three Mile Island Public Health Fund decided in 1985 to undertake the preparation of its own plan, based upon the state of the art in emergency planning, for the TMI nuclear plant. After circulating a request for proposals, the Fund awarded a contract to a team of researchers based at Clark University in Worcester, Massachusetts.

To initiate their study, the researchers first conducted an appraisal of the adequacy of the existing emergency plan for the Three Mile Island nuclear plant and, more generally, the state of emergency preparedness around nuclear power plants in the United States. This assessment revealed very serious generic shortcomings, many of them identified by those most actively involved in the implementation of emergency plans. The objectives that currently drive emergency preparedness are vague and fail to provide sufficient guidance to those responsible for plan development and implementation. The plans do not have the necessary capability to anticipate severe accident conditions and, when appropriate, to initiate early precautionary responses, before an accident is fully under way. A particular concern is the rigidity that reflects a heavy reliance upon command-and-control operations and "pro forma" adherence to federal criteria and guides. The plans tend to be weak in arrangements for emergency medical response and post-accident recovery. Moreover, they cover too small a geographical area given the potential for distant dispersion of radioactive material such as happened at Chernobyl.

Often the emergency plans rest upon assumptions about public coping and behavior that are inconsistent with the knowledge that has evolved since the early 1980s. Although drills and exercises are now commonplace, they are frequently unrealistic and fail to provide either training or the means for validation of the plan. Finally, when emergency plans are found wanting, current regulations lack "teeth" to force prompt and full rectification. Taken together, these deficiencies cast doubt on the adequacy of present emergency plans to protect the public during a serious nuclear accident and make it clear that a fresh approach to emergency preparedness is needed. The current effort by the Nuclear Regulatory Commission to "streamline" the licensing process could well exacerbate these problems.

The plan that follows is the product of a multi-year effort to develop a plan specific to the needs of the TMI region, taking account of the existing TMI plans and the current state of knowledge in emergency planning. The plan, however, is directly relevant to emergencies at nuclear power plants more generally and seeks to address generic issues. It is a model plan that can inform by example. Since the authors have had the mandate to develop the best plan possible regardless of existing federal regulations and other constraints, this plan offers both public officials and the public at large an alternative and more ambitious approach to emergency preparedness. It is, in this sense, a conceptual plan. It contains the essential core of emergency preparedness but would, if adopted, require the addition of procedures for implementation.

These detailed procedures are best developed by the individuals and organizations responsible for implementing emergency planning and preparedness arrangements. Such details are beyond the scope of this effort, and would detract from the clarity of the conceptual plan.

Instead, the authors have chosen to convey the essence of effective and comprehensive planning in an integrated and highly readable document, while recognizing that adoption of the plan would require detailed procedures for implementation. Thus, references and footnotes do not appear in this plan; the technical supporting papers in Volume 2, however, do note at length the relevant literature and evidence. Although more ambitious than existing plans, this plan is not based on "pie-in-the-sky" attempts to use every technology or cover every contingency. Detailed planning for implementation is needed to determine the additional resources and costs required. Yet our estimate is that the added costs, although substantial, are not high compared to present expenditures on emergency planning. The most substantive change called for is a much greater institutional commitment that sees emergency planning as one of the essential safety systems for nuclear plants.

Individuals from the TMI region and from federal, state, and local government were consulted in the development of this model plan. An actual plan to be implemented, however, should be a much more collaborative effort, involving extensive public participation. Emergency planning, in the view of the authors, is a collective process that depends centrally upon the adaptability and capabilities of public officials, emergency workers, and individual citizens.

Two principles have guided the preparation of this model plan. First, people are more likely to respond reasonably and effectively if they are provided full information during an emergency, if they understand why particular actions are necessary, and if well-considered plans and supporting resources are in place. Second, an effective plan must seek to capitalize on the problem-solving abilities and special knowledge of individuals and groups in emergency situations, thereby ensuring flexibility, resiliency, and efficiency in emergency response. Volume 2 of this study provides supporting analyses and documentation for these positions.

This plan specifies the criteria, decisions, actions, and resources necessary to achieve, in the event of an accident, a high level of preparedness and an effective emergency response. As with existing plans, this conceptual plan states that certain decisions or actions "will" occur and that certain resources "are" or "will be" available. The particular resources and the authority for particular decisions and actions may not now be in existence, but they would be required were the plan to be implemented.

THE MODEL PLAN IN CONTEXT

1. How did emergency planning develop?

The current state of emergency planning has evolved as follows:

- Early federal standards on reactor siting were conservative, and if applied to TMI, would not permit people to live within 16 miles of the plant.
- Siting standards were relaxed during the 1950s, on the assumptions that severe reactor accidents were not credible and that containment buildings would prevent large releases.
- Awareness grew during the 1970s that severe accidents might after all be credible, and the current theoretical basis for off-site emergency planning was developed just before the 1979 TMI accident.
- The existing emergency-planning regulations, adopted in the aftermath of the TMI accident, initially gave an important role to state and local governments.
- In an effort to override those state governments that argue that adequate emergency planning cannot be undertaken for plants such as Shoreham and Seabrook, the federal government has recently amended its regulations to allow utilities to conduct off-site emergency planning with federal assistance.

Early guidance on nuclear-reactor siting. In 1950, the US Atomic Energy Commission (AEC) issued guidance for siting the small research reactors of the day. This guidance called for an "exclusion area" surrounding the reactor. Application of that guidance to the TMI plant would dictate an exclusion area with a 16-mile radius, within which people could not live. The AEC relaxed its standard during the 1950s, however, on the twin assumptions that severe (core melt) accidents were not credible and that containment buildings would prevent large releases from lesser accidents. Thus, the exclusion area for the TMI plant lies entirely within the banks of the Susquehanna River.

Emergency planning before the TMI accident. During the 1960s, the prevailing view was that reactor accidents leading to significant on-site or off-site radiation exposure were sufficiently unlikely that emergency planning deserved only a low priority. The AEC did issue regulations in 1970, however, requiring on-site emergency planning. Then, in 1973 the Office of Emergency Preparedness, a predecessor of the present Federal Emergency Management Agency (FEMA), called upon the AEC to provide off-site emergency planning assistance to state and local governments. Thereafter the AEC and the US Nuclear Regulatory Commission (NRC), its regulatory successor since 1975, implemented an essentially voluntary program of planning assistance to state governments. In 1976, the NRC also established, jointly with the US Environmental Protection Agency (EPA), a task force to establish a planning basis for radiological emergencies. This group took partial account of the possibility of core-melt accidents with containment failure, as identified in the NRC's 1975 *Reactor Safety Study*. Their report was published in December 1978.

Emergency planning after the TMI accident. The March 1979 TMI accident demonstrated that core-melt accidents are indeed credible. In response to recommendations made by a presidential commission he had established, President Carter called in December 1979 for an upgrading of off-site emergency planning, and he assigned FEMA the responsibility of overseeing this process. The NRC, EPA, and FEMA accepted the planning basis previously proposed by the NRC/EPA task force, and this theoretical basis was incorporated into NRC emergency planning regulations, which took effect in August 1980. Most prominently, those regulations embody the task force's recommendation that plans for evacuation and sheltering encompass a zone with a 10-mile radius.

Involvement of state and local governments. Nuclear Regulatory Commission regulations initially stipulated that state and local government emergency plans, certified as adequate by FEMA, must be in place before a nuclear plant operating license is issued. In principle, the continued operation of previously licensed plants is subject to the same standard, although enforcement has been more lax in such cases. Some state and local governments have concluded that adequate emergency planning cannot be undertaken for certain plants. Notably, the governments of New York and Massachusetts have declined to put forward emergency plans for the Shoreham and Seabrook nuclear plants respectively, thus delaying the granting of the Seabrook operating license and forcing the abandonment of Shoreham. In an attempt to override these state objections, the NRC has amended its regulations to allow utilities to conduct off-site emergency planning. Moreover, by an Executive Order issued in November 1988, President Reagan authorized

FEMA to assist utilities in making such plans and, during an actual emergency, to coordinate federal government responses that substitute for state and local government actions. The effect of this action has been to undermine the essential recognition that successful emergency response depends centrally upon the active involvement of local officials and publics in planning for their own protection. By contrast, this model plan for nuclear power plants moves in the opposite direction—toward preparedness predicated upon local knowledge and circumstances.

2. How does this plan differ from existing plans?

Although emergency plans for nuclear power plants differ significantly from utility to utility and sometimes from site to site, they share many common features due to the US Nuclear Regulatory Commission guidelines that they all must meet. The proposed plan that follows did not begin with these regulatory requirements as an assumption, however, but rather with the needs for emergency preparedness suggested by the current understanding of nuclear power plant accidents, emergency decision making, and the response of publics and special groups to accident situations. It is not surprising, then, that this plan differs significantly from existing plans in important respects. Table 1 contrasts the major areas of difference.

TABLE 1

Existing Versus Proposed Planning Arrangements

	Existing	Proposed
Objectives	Unstructured and nonspecific: "Dose Saving and in some cases life-saving"	A specific hierarchy of planning goals - life and early injury saving - dose reduction below Protective Action Guides (PAGs)
Flexibility and Resiliency	Strong centralization of authority and use of "command-and-control" operations Emphasis on programming emergency behavior Redundancy in communications system	Increased role for county and local governments Emphasis on local knowledge and adaptive behavior of emergency workers and publics More information-rich emergency environments
Precautionary Initiation of Emergency Response	Federal regulations call for initiation of response if core melt has occurred or is "imminent" Recognition on the part of some state and federal officials of the merit of early initiation of response, but a lack of clear decision criteria A lack, among state authorities, of technical capability to identify and evaluate plant conditions that call for early initiation of response	Precautionary initiation of response when probability of core melt becomes significant (10 percent or greater) State authorities to have the technical capability promptly to identify and evaluate plant conditions that call for precautionary initiation of response

Criteria for Long-term Protective Actions	Protective Action Recommendation (PARs) to be based on measurements of contamination levels of food, water, land, and buildings Absence of comprehensive and consistent guidelines regarding significant long-term exposure	PARs to be based on measurements that are compared with pre-established long-term guidelines for annual exposures—set as a small fraction (I/10) of the short-term Protective Action Guides (PAGs)
Planning Zones and Protective-Action Options	Two Zones: 0-10 Miles: Plume Protection—evacuation and/or sheltering when core melt is imminent or release is in progress 0-50 Miles: Ingestion Protection	Three Zones: 0-5 Miles: Plume Protection—emphasizing early* evacuation 5-25 Miles: Plume Protection—evacuation and sheltering based on weather conditions and monitoring >25 Miles: Plume Protection—sheltering and evacuation or relocation based on monitoring All distances: Ingestion Protection
Accident Classification	Four classification levels based partly on severity of plant condition, partly on size of release; accident probability not built into definitions of classification levels.	Four classification levels based on likelihood of core melt, plus one level for accidental releases without major core damage

(continues)

* when core melt is only a possibility

TABLE 1 *(continued)*

	Existing	Proposed
Increased State Capabilities and Power	Utility responsible for identifying an emergency and classifying its severity	State authorities to have the technical capability to identify and classify independently the severity of an emergency
	Heavy reliance by state authorities on the utility for information about plant conditions during an emergency	State authorities allowed to classify an emergency at a higher level of severity than proposed by the utility
	State authorities' limited capability for plume tracking and dose projection	State authorities to have a quick-reaction capability for plume tracking and dose projection
Local Government	Risk counties (five in the TMI area) responsible for coordination of development and implementation of emergency plans	All counties (seven in the TMI area) potentially at risk coordinate development and implementation of emergency plans
	Limited "horizontal" communications among counties and among municipalities	Extensive "horizontal" communications among counties and among municipalities
	Limited training, especially for volunteer emergency workers and staff at schools, hospitals, and other institutions	Extensive training, especially for volunteer emergency workers and staff at schools, hospitals, and other institutions

Monitoring and Plume Tracking	State authorities and limited capability of utilities for plume tracking and dose projection Federal authorities' substantial plume tracking and radiation monitoring capability subject to several hours' delay before initial deployment Relatively crude systems for transmission and processing of monitoring data, even at the federal level	State authorities and the utility to have quick-reaction airborne monitoring and plume tracking capability, and enhanced ground-based monitoring capability Monitoring data to be transmitted and processed in real time Monitoring to be integrated with a joint state-federal computer modelling capability, allowing prompt dose projection Radiation monitoring and dose projection information to be promptly provided to TV stations in a form suitable for direct display
Medical Response	Plans for screening and external decontamination of up to 20% of people in 10 mile zone Arrangements for treating a small number of seriously exposed people No explicit planning for integration of national medical resources in emergency	Plans for screening and external decontamination of persons potentially exposed or concerned about potential exposure Plans to call in national medical resources in the event of a severe accident in which hundreds and thousands of people may require treatment for serious radiation exposures

(continues)

TABLE 1 *(continued)*

	Existing	Proposed
Information Systems	Off-site authorities dependent on utility for information on plant conditions	Off-site authorities to have independent capability to gather information about plant conditions
	Overly centralized communications network with preponderance of "vertical" links	Enhanced "horizontal" communications, increased redundancy, and decentralization
	Public alert and notification reliant on: - fixed sirens - route alerting	Enhanced alert and notification using: - automated dialing system - tone alert radios - fixed sirens - route alerting
	Media transmission of Emergency Broadcast System (EBS) messages and other information obtained from multiple sources and media briefings	On-line media access for EBS messages and a single source of official information
Drills, Exercises, and Training	Limited training especially for volunteer emergency workers and staff at schools, prisons, nursing homes, and hospitals	Extensive and comprehensive training for all emergency workers
	Review of planning arrangements by State, FEMA, and NRC	Additional review by newly created Local Review Committee
	Routinized, unrealistic, and limited exercises necessary to maintain compliance, lack of which does not lead to suspension of license to operate	Maintenance of the operating license conditional on successful evaluation in rigorous, realistic, and comprehensive exercises

Public Education	Pre-emergency education centering upon dissemination of brochures by utility as well as emergency procedures distributed to key locations Emergency information emphasizes alert and notification by Governor and state emergency-management agency (PEMA at TMI)	Intensive, broad-based pre-emergency educational program in Inner Planning Zone: Information dissemination in all zones Citizen involvement in design and implementation of educational program Upgraded information flow during emergencies
Treating Emergency Preparedness as a Plant Safety System	Federal requirement of an emergency plan in place as a condition for continued plant operation No requirement that the plan be capable of execution at all times when the plant is operating	Emergency preparedness regarded as a safety system equivalent to in-plant system Significant degradation in the state of emergency preparedness (e.g., blockage of roads by snow) to be grounds for shutting down the plant The plant's operating license to be contingent on the continuous maintenance of an effective emergency-response capability

OBJECTIVES

3. What are the objectives of the plan?

Establishing clear objectives is a fundamental requirement for effective emergency planning and preparedness.

The objectives of this plan are:

1. to minimize the harm to life associated with a range of nuclear power plant accidents, encompassing both minor events and the rarer, more severe accidents. More specifically, the plan seeks to:
   a. provide an enhanced capability for avoiding radiation injuries even in very severe accidents;
   b. keep any radiation doses to the public below the Protective Action Guides (see Section 6 below); and,
   c. reduce any individual doses to levels as low as reasonably achievable below these Guides and minimize the collective dose to the population.
2. to make the implementation of any protective actions as equitable and as cost-effective as reasonable.
3. to educate the public about the risks of nuclear power plants and the means by which people can protect themselves.
4. to capitalize as much as possible on local knowledge and human ingenuity.
5. to build (and merit) public trust and confidence in the emergency warning and response system.
6. to provide guidance and plans for the allocation and utilization of resources to promote long-term recovery following a major accident.

These objectives, though all desirable, are not of equal importance. They are also sometimes in conflict. It is unlikely, for example, that the emergency plan that is most economical will also be the one that provides greatest fairness in protecting all potentially exposed people—those near the plant and those far away, children at school and parents at work or home, those with cars and those who rely on public transportation.

The primary objective is the first one—to minimize the harm to life from any accident. The other objectives—such as educating the public and capitalizing on local knowledge—are intended either to help accomplish this goal or to assure that the plan is realistic and fair. How to weight the secondary objectives (2-6) is necessarily a matter for the discretion and judgment of the public officials who will select the recommended protective actions during an emergency situation.

Beyond these objectives, an effective emergency plan should have other desirable characteristics. A good emergency plan should:

- be *clear, concise,* and *easily understandable* by emergency workers and the public;
- be *flexible* and *resilient* (i.e., maintain a basic emergency-response capability in the face of failures);
- support institutional roles and responsibilities with the appropriate *allocation of resources and authority;*
- be *comprehensive* (i.e., address the full range of potential accidents, the spectrum of available protective actions, and contingencies such as adverse weather, floods, and traffic conditions);
- be *realistic* in assumptions (especially in regard to human behavior); and,
- be *coherent, consistent,* and *unambiguous.*

NUCLEAR ACCIDENTS
AND THEIR EFFECTS

---

4. What is a nuclear emergency?

Nuclear power plants routinely release very small amounts of radioactivity. In addition, they often experience minor interruptions in their operation. More rarely, they experience events that have the potential to lead to a release of radioactivity sufficient to be of public-health concern.

In this plan, the word "emergency" refers to an event at the plant that releases, or threatens to release, a significant amount of radioactivity to the atmosphere. The word "accident" is also used throughout this plan. Although common usage dictates that the word be used for events of widely varying significance, the accidents of interest here are those that are also emergencies. For purposes of emergency planning, the following properties of nuclear plants are particularly important:

- nuclear plants generate large amounts of radioactive materials as a by-product of normal operation;
- accidents can occur, however, in which much larger amounts of radioactive materials can be released to the atmosphere as gases or tiny particles; and,
- the nature of the radioactive release to the environment during an accident can vary widely and, even for a particular type of accident, is predictable only within wide bounds.

The TMI plant. The site at Three Mile Island contains two nuclear power plants—Unit 1 and Unit 2. Both units are pressurized water reactors (PWRs), the most commonly used type of commercial reactor in the world. TMI Unit 2, severely damaged in an accident in 1979, will not resume operations.

Figure 1 is a schematic drawing of TMI Unit 1. At the heart of the plant is a reactor core, composed of many long, thin fuel rods in a cylindrical bundle, inside a steel pressure vessel. Cooling water is pumped

FIGURE 1

Pressurized Water Reactor Power Plant
(schematic drawing)

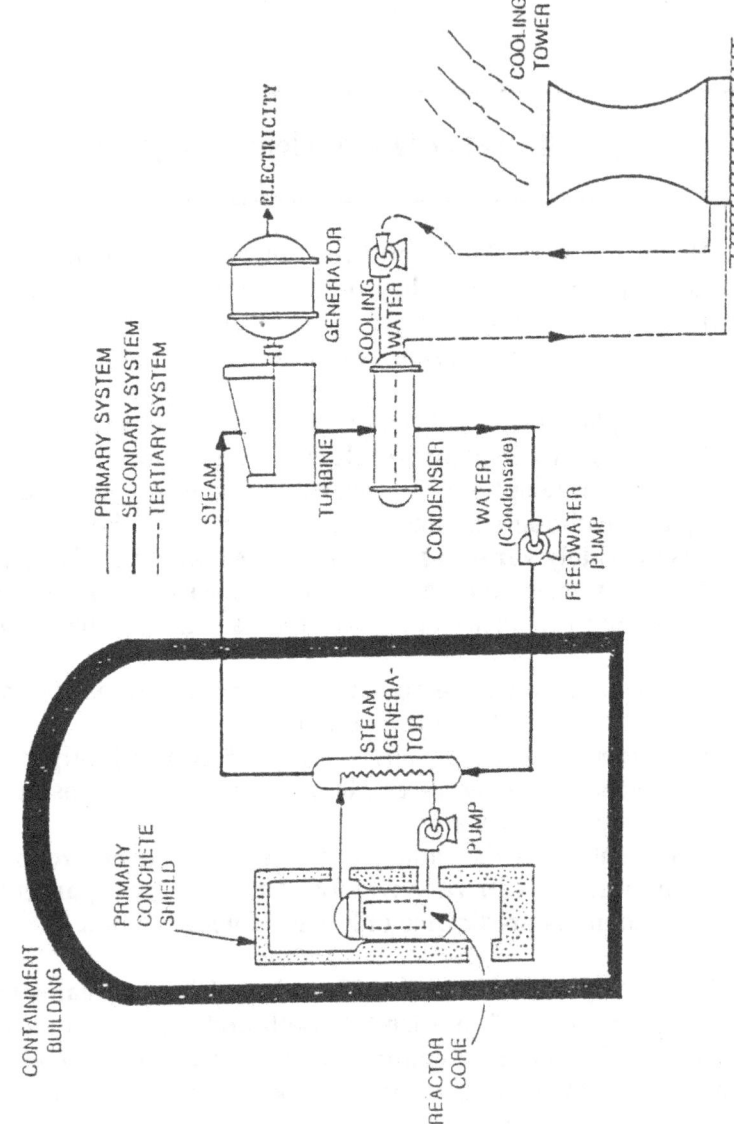

through this vessel, carrying heat to the steam generators, in which water turned to steam then turns a turbine coupled to a generator, thus producing electrical power.

A large containment building of reinforced concrete houses the reactor pressure vessel and the steam generators. This building, designed to contain the radioactivity released from the core during certain types of accident, will not necessarily withstand the effects of an accident during which a large part of the core melts.

Potential emergencies at TMI. Over future years, the TMI plant is likely to experience events that are potential emergencies. More rare may be events with a significant probability of leading to a release substantially larger than routine releases. Still more rare may be events that actually lead to such a release. The latter two classes of events will be true "emergencies," as shown in Figure 2.

Releases may arise from a core-melt accident, from an accident not involving core melt, or from a spent-fuel storage accident (see Figure 2). Within this plan, the focus is on core-melt accidents, because preparations for such accidents must be more extensive than those for the other two classes of release.

A core melt can occur if the reactor's supply of cooling water is interrupted for more than a short period. Such an accident is particularly serious because it will liberate the radioactive materials contained in the irradiated fuel. During the 1979 accident at TMI Unit 2, partial melting of the reactor core did occur.

In a core-melt accident, radioactive materials released from the core can reach the outside atmosphere through three basic routes:

- directly from the interior of the reactor vessel, through pipes passing through the containment wall, entering the atmosphere directly or through the auxiliary building;
- from the reactor vessel into the containment building and then into the atmosphere, directly or via adjacent buildings; or,
- through the molten core's penetration of the floor of the containment building, thus allowing radioactivity to reach the atmosphere through the surrounding soil.

Accidents of a variety of types at TMI could release radioactive materials into the Susquehanna River, with or without an accompanying atmospheric release. Specialized emergency responses would be necessary in the event of river contamination. In this plan, the focus is on atmospheric release.

FIGURE 2

Types of Nuclear Emergency

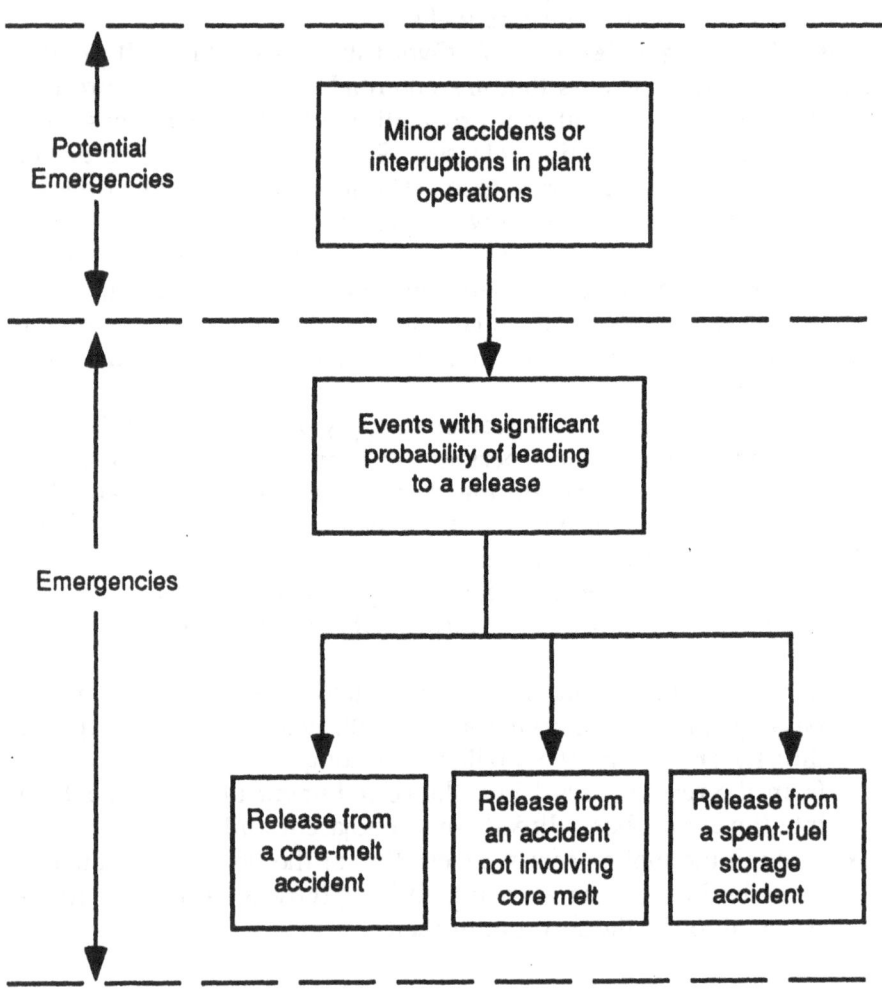

The timing of a core-melt accident. After an accident has begun, a delay—ranging from as little as 15 minutes to as much as 12 hours—will occur before the reactor core begins to melt (see Figure 3). Once core melting proceeds, the potential exists for an atmospheric release, which may, however, take as much as a day to occur.

When core melt is completed, the molten core will slump to the base of the reactor vessel. It will then melt through the vessel's lower wall. This represents a "crisis point" in the progression of the accident. At this point, several phenomena—especially steam explosion, high-pressure melt ejection, and hydrogen explosion—could lead to rapid failure of the containment. If this were to happen, a large, fast-breaking release could occur. If no release occurs before or during this critical period, a large release is relatively unlikely for the next few hours.

Once a release has begun, it may last for less than an hour, or it may last for several days. The composition and magnitude of the release may change during the release period. There may, for example, be "surges" in the rate of release. Some potential for release may persist for weeks after the initial accident.

The radioactive plume. Many different radioactive materials are present in the reactor core. They represent a substantial hazard if released to the environment.

Many types of accidents are possible, and each will have its unique release (see, for example, the Box on the Chernobyl accident). The

---

THE RELEVANCE OF THE CHERNOBYL ACCIDENT

In 1986, an accident involving a "power excursion" occurred at Unit 4 of the Chernobyl nuclear station in the Soviet Union. As a result of design deficiencies and operator errors, the reactor's power output rapidly rose to a level hundreds of times the normal operating level. This led to a violent explosion and subsequent fire. Due to differences in design, pressurized water reactors (PWRs), such as those at the TMI plant, are not expected to exhibit power excursions of this type. Such reactors may, however, experience accidents that—like the Chernobyl accident—lead to a release to the atmosphere of a substantial part of the core inventory of radioactive materials.

The radioactive plume from Chernobyl rose high into the atmosphere. If the same release had occurred without that vigorous plume rise, radiation doses nearby would have been substantially greater. Some severe PWR accidents can produce a high plume rise like that at Chernobyl, whereas others may have little or no plume rise.

FIGURE 3

The Timing of a Core-Melt Accident

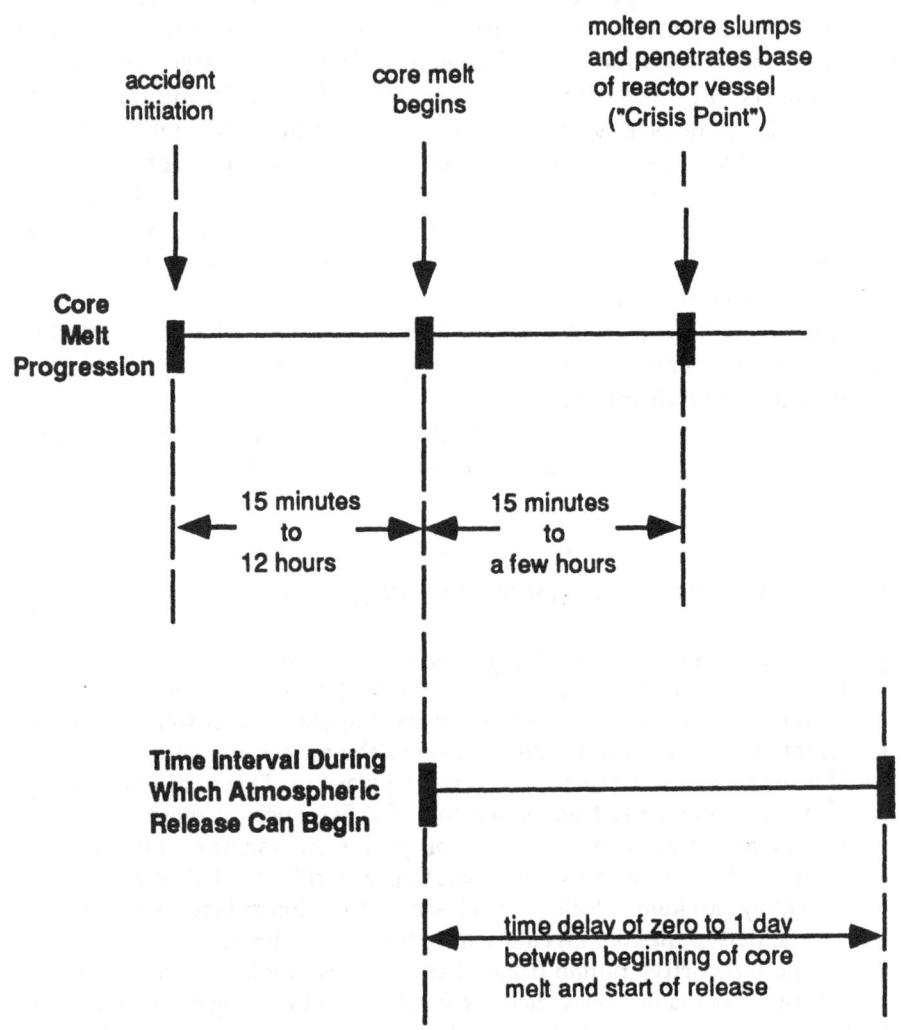

FIGURE 4

Examples of Plume Types

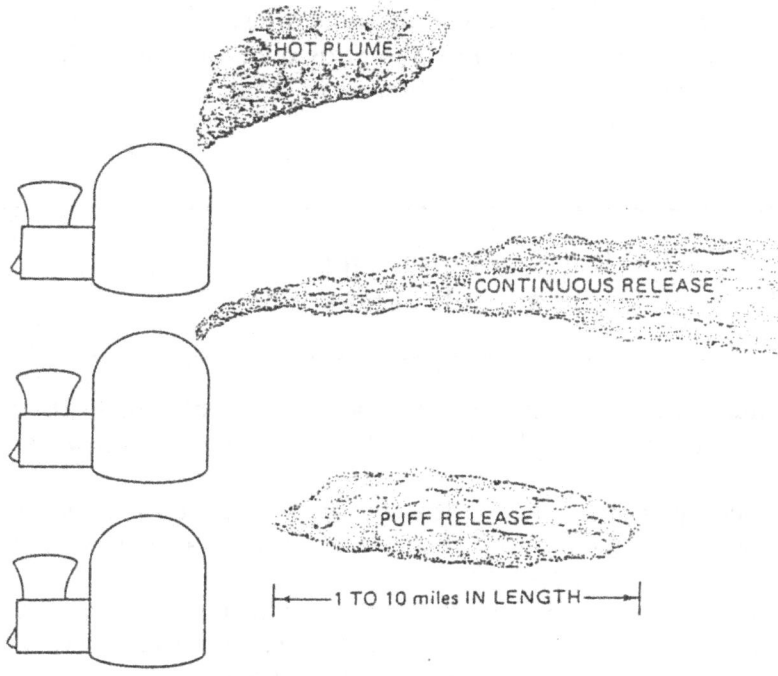

fraction of the core inventory of a given radioactive material that reaches the atmosphere could be very small or could be as high as tens of percent. Although a considerable research effort has been conducted over the last two decades, it is still not possible to predict accurately the magnitude of a release, even for a particular type of accident.

The radioactive plume could be released from the containment building or from an adjacent building. In some cases, the plume will contain considerable heat, and will rise due to its own buoyancy. The plume may be released over many hours, or in a relatively brief puff (see Figure 4).

---

5. What kinds of radiation exposure may occur as a result of an accident?

A nuclear accident that releases radioactive materials to the atmosphere may expose people to radioactivity in four principal ways:

- people may inhale radioactive materials, transported by the wind as an invisible cloud, or be exposed to penetrating radiation as the cloud passes;
- people may be exposed to penetrating (gamma, beta, and x) radiation from radioactive materials that are deposited on the ground and other surfaces as the cloud moves past;
- people may eat foods or drink water or milk contaminated with radioactivity; and,
- people may be exposed by absorption of radioactive substances (such as Iodine-131) through the skin and by the deposition of radioactive particles on the skin.

Dispersion of the radioactive plume. Winds will carry the radioactive plume away from the nuclear plant. The concentration of radioactivity will generally decrease as the plume moves farther away from the plant. The pattern of dispersion of radioactive materials will be affected in the following ways:

- high wind speeds and turbulent atmospheric conditions (as during a clear sunny day) will lead to rapid dispersal of the plume;

- low wind speeds and very stable atmospheric conditions (as during a clear night) will lead to a narrow plume with a relatively high concentration of radioactivity;
- a hot plume will rise quickly and prevent high concentrations of radioactivity for the first few miles downwind but may contaminate extensive areas; and,
- rainfall, fog, or other precipitation may cause concentrated pockets of radiation, or "hot spots," out to considerable distances from the plant (see Box).

Properties of radioactive materials in the plume. Some materials (such as the element krypton) will always be in gaseous form. Other materials (such as the element cesium or its compounds) will be in the form of tiny particles that adhere strongly to surfaces such as soil, buildings, and clothing. Radiation from these particles can lead to human exposure after the plume has passed. Still other materials may, if inhaled, concentrate in a particular body organ (the thyroid gland, in the case of iodine). Each radioactive element will be present in one or more species (isotopes), each with a characteristic type of radiation and rate of decay. Over periods ranging from hours to

---

THE EFFECT OF RAIN, SNOW, OR FOG

If the radioactive cloud encounters rain, snow, or fog, radioactive materials may fall to the earth's surface in much greater quantities than would otherwise be the case. This will, in turn, lead to greater deposition of radioactivity on the ground and on other surfaces and may increase the exposure of people to radiation. "Hot spots" of this kind can arise at any distance: within a few miles of the plant or even hundreds of miles away.

The 1986 Chernobyl accident demonstrated the importance of rainfall. At any given distance from Chernobyl, places with rainfall had 15-20 times as much radioactivity as dry places. Rainfall locations in southern Germany and central Austria, 600 or more miles from Chernobyl, had levels of radioactivity comparable to those arising from dry deposition at about 100 miles from the plant.

In the event of an accident at a nuclear plant, radiation monitoring teams will need to pay special attention to areas of precipitation. High radiation levels at such locations should command appropriate emergency responses.

decades or longer, the intensity of the radiation emitted by these isotopes will fall to low levels.

Types of radiation. Materials in the plume can give out three types of radiation:

- *alpha radiation,* which, due to its short range in air or tissue, is of concern only if alpha-emitting material is inhaled or ingested;
- *beta radiation,* which is somewhat more penetrating, is also of concern if beta-emitting material is inhaled or ingested, and can in addition cause skin and subcutaneous damage if the material is on or close to the skin, as well as damage to the lens of the eye and to testicles; and,
- *gamma radiation,* which is very penetrating and can travel great distances, so that radiation from a passing radioactive cloud or from material deposited on the ground may reach all organs of the human body.

Pathways for human exposure. A person may be exposed to radioactivity via one or more of the pathways shown in Figure 5. Cloud shine (penetrating radiation from the passing plume) and inhalation will be of concern only during the period of plume passage. Contamination of skin and clothing and exposure to contaminated ground will be of concern pending the decontamination and evacuation of people from affected areas. The ingestion pathway will usually be of less immediate concern.

Radiation doses to the body. Various parts of the human body—such as the bone marrow, thyroid gland, or lungs—will be exposed to a different type and level of radiation. Gamma radiation from the passing cloud or from materials deposited on the ground will irradiate organs more or less uniformly, but each inhaled or ingested radioactive element will be distributed to body organs in a different way.

The radiation dose to an organ is expressed in *rem,* or sometimes in *millirem* (where 1,000 millirem = 1 rem).* Residents of Pennsylvania typically receive a dose to their average body organ of about 100 millirem of radiation each year from natural sources such as cosmic radiation and the radioactivity in building materials, and about 100 more millirem from man-made sources such as medical and dental x-rays.

---

*An alternative dose unit is the Sievert (Sv), which equals 100 rem, and millisievert (mSv), which equals 0.1 rem.

27

FIGURE 5

Radiation-Dose Pathways

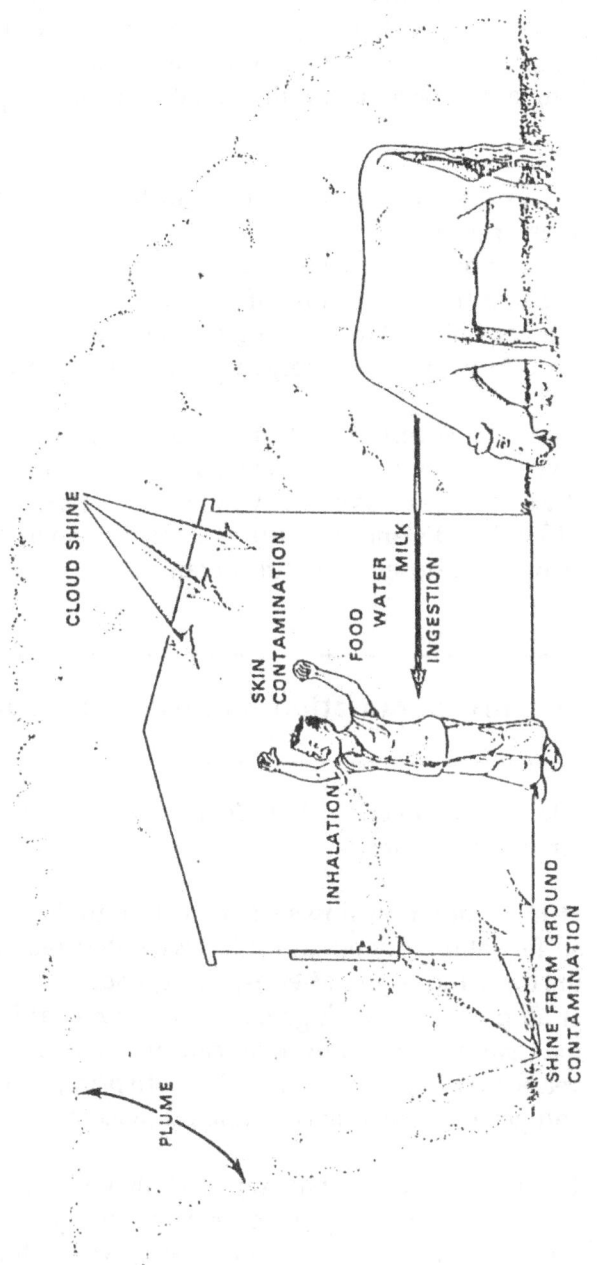

Inhalation of naturally occurring radon gas (strongly discharged from the soil in parts of Pennsylvania) can give an additional dose to the lungs of as much as 500-5000 millirem annually.

During a reactor accident, the radiation dose to a person will depend upon:

- the inventory of radionuclides in the reactor and the fraction of these released;
- the size of the release and extent of plume rise;
- the weather at the time of release;
- the person's distance from the reactor; and,
- the effectiveness of emergency-response actions.

For the most severe (worst-case) accidents, a person downwind who takes no emergency-response actions could suffer—during the first few days of exposure—an average (whole-body) dose of up to hundreds of rem at 10 miles' distance and up to about 10 rem at 100 miles' distance, and much higher doses to specific organs.

---

6. How is radiation harmful to human beings?

People who are exposed to radioactivity may be at risk of two types of adverse health effects:

- at high radiation doses—tens or hundreds of rem—people may suffer fatal or debilitating illnesses that manifest themselves in the days and weeks after exposure; and,
- at all doses, but with generally lower risk at lower doses, people may suffer potentially fatal cancers, which manifest themselves years after exposure. Also, their offspring may suffer teratogenic and genetic defects and childhood cancers.

Early health effects. At average (whole-body) doses of tens or hundreds of rem, people may experience a variety of health effects that manifest themselves over the days and weeks following exposure. These "early" health effects include death and a variety of types of debilitating but nonfatal illnesses (see Table 2).

Early health effects may be associated with symptoms that vary over time. For example, acute irradiation of the whole body will lead

TABLE 2

**Early Health Effects from Acute Radiation Exposure:
Dose Required to Induce Effect
in Most Sensitive 50% of Population**

Health Effect	Dose (rem) to the relevant organ(s)
DEATH	
bone-marrow damage	350
lung damage:	
adult	950
child	480
gastrointestinal damage	1,500
prenatal/neonatal death	100
ILLNESS	
vomiting	180
impaired lung function	400
cataracts	310
sterility:	
female (permanent)	260
male (temporary)	70
microcephaly (fetal exposure)	40

to initial symptoms, such as vomiting, followed by a period of two-three weeks without symptoms before onset of the main illness (hemorrhage, hair loss, susceptibility to infections), which might last several weeks and result in death.

Most of the early health effects have dose thresholds, below which the effect is not apparent. Also, adverse health effects generally are less likely if the dose is spread over many days, rather than being concentrated in the first day.

Some people are more likely to suffer early health effects from a given exposure to radiation. Healthy young adults have the greatest resistance, whereas sick people, children, and the elderly have less resistance. Fetuses are particularly sensitive, and may die or develop gross developmental abnormalities at doses at which adults would show no effects.

Intensive medical care can improve the chances of survival for people who are heavily exposed. For example, about 500 people were hospitalized in the Soviet Union after the 1986 Chernobyl accident. Of these, about 50 people had average (whole-body) doses exceeding 500 rem. Medical treatment (such as blood transfusions, locating of patients in sterile environments, and use of anti-infective drugs) allowed patients to survive at doses about one-third higher than would otherwise have been the case.

Delayed health effects. Years after their exposure to radiation, people may suffer cancers related to the exposure. There are many causes of cancer, however, and it may not be possible to identify which cancers might have been caused by radiation exposures, and which cancers might have resulted from other causes.

Considerable scientific uncertainty exists about the relationship between dose and the incidence of cancers. It is generally assumed, however, that the risk of cancer is greater at higher radiation doses, but—in contrast to most early health effects—there is no dose threshold. In other words, people with high doses of radiation are more likely to develop cancer than those with lower doses, but even those with extremely low doses have some increased risk. As a guide, one can estimate that one excess cancer will occur per 1,000 people each exposed to a 1-rem average (whole-body) dose, with proportionately more (or fewer) excess cancers at higher (or lower) doses.

Different populations have different susceptibilities to cancer, and some individuals are more sensitive due to hereditary traits. At the same radiation exposure, women generally experience somewhat more radiation-induced fatal cancers than men, largely due to the added risk of cancer of the breast. Children are at greater risk of subsequent cancer than adults.

Genetic effects may occur in the offspring of irradiated populations. Again, considerable scientific uncertainty exists about the relationship between dose and the incidence of genetic effects. According to the US National Research Council, an exposure of 1 rem per person throughout the US population would increase the number of serious genetic disorders per million children by up to 50 cases, with proportionately more or fewer cases for higher or lower doses. For comparison, about 10 percent of children currently suffer genetic disorders of varying degrees of seriousness.

Protective action guides (PAGs). The preceding discussion demonstrates that there is no "safe" dose of radiation. Thus emergency planning seeks, ideally, to eliminate radiation exposures. In practice, this will not be possible, and so emergency planning seeks to keep radiation doses below particular levels. These levels are known as *protective action guides* (see Box).

PROTECTIVE ACTION GUIDES

Criteria needed to determine when particular protective actions are required may be expressed as projected radiation doses. Thus, if some people are likely to receive a radiation dose in excess of the dose criteria—the Protective Action Guides (PAGs)—then protective actions are imperative. Whenever possible, actions should be taken to reduce doses even further, and no dose that can easily be avoided should be regarded as acceptable.

In this plan, different dose criteria guide short-term and long-term protective actions (for the distinction between these two types of action, see Section 18). This reflects two considerations: first, that there is more potential for dose-reducing actions in the longer term; and, second, that many more people may be exposed in the longer term.

Protective Action Guides cannot be objectively determined and should therefore be set and periodically reviewed within the arena of public policy. In this plan, the Short-Term Protective Action Guides are those currently recommended by the US Environmental Protection Agency (EPA), whereas the Long-Term Protective Action Guides are expressed as an annual dose one-tenth as high as the Short-Term Protective Action Guides. We recommend that these criteria be reexamined with public comment every five years.

Following the EPA, we list Short-Term Protective Action Guides for two types of radiation dose—to the whole body and to the thyroid—as shown in Tables 3 and 4 .

TABLE 3

**Short-Term Protective Action Guides
for Whole-Body Exposure to a Radioactive Plume
or to Deposited Radioactivity**

Population at Risk	Projected Whole-Body Gamma Dose (in REM)
General population	1 to 5*
Emergency workers	25
Emergency workers involved in life-saving activities	75

*When ranges are shown, the lowest value should be used, especially for sensitive populations and if there are no major local constraints in providing protection at the level. In no case should the higher value be exceeded in determining the need for protective action.

TABLE 4

Short-term Protective Action Guides
for Thyroid Dose Due to Inhalation
from a Passing Plume

Population at Risk	Projected Thyroid Dose (in REM)
General population	5-25*
Emergency workers	125
Emergency workers involved life-saving activities	**

*When ranges are shown, the lowest value should be used, especially for sensitive populations and if there are no major local constraints in providing protection at the level. In no case should the higher value be exceeded in determining the need for protective action.

**No specific upper limit is given for thyroid exposure since in the extreme case complete thyroid loss might be an acceptable penalty for a life saved. This should not be necessary, however, if respirators and/or thyroid protection for rescue personnel are available as the result of adequate planning.

---

7. Will an emergency plan assure safety?

Emergency planning commands an important but restricted role in protecting the safety of those living around nuclear plants. This role includes:

- providing the last layer of safety for a nuclear plant and reducing, to the extent possible, the harm from a nuclear plant accident;
- eliminating, under favorable conditions, most "early health effects" and reducing the incidence of delayed health effects; and
- reducing, in limited ways, the long-term public health problems associated with a radioactively contaminated environment.

At the same time, it must be noted that emergency planning if poorly done will do little to reduce, and could even enlarge, the adverse consequences of a nuclear accident.

Layers of safety for a nuclear plant. Nuclear power plants pose a significant potential hazard because they concentrate a large inventory of hazardous material in a small space—the reactor core. People are protected from this potential danger by five layers of safety:

- plant design
- quality of construction
- quality of operation
- selection of the site
- emergency planning and preparedness

Plant design is a fundamental layer of safety. Unless the design is basically sound, safety cannot be assured no matter how carefully the plant is built and operated.

Two other layers of safety are the quality of construction and the quality of operation. In these areas, the recruitment and training of a highly qualified workforce with an uncompromised commitment to safety are essential. An accident leading to a radioactive release represents a weakness in plant design or a failure in one or both of these other two layers.

Given a release, the public may gain some protection from the remote siting of the plant. At many US nuclear plants, however, large numbers of people live relatively close to the plant.

Emergency planning is a fifth layer of safety. Emergency preparedness, however, needs to occur well in advance of any potential accident. When a release has occurred, it is the last line of defense in protecting people from exposure to radioactivity.

The contribution of emergency planning and preparedness. A variety of consequences may arise from an accident and a radioactive release. Emergency preparedness can reduce the scale of some, but not all, of these consequences (see Table 5).

Early health effects are most likely to occur within a radius of five miles from a nuclear plant. Certain weather conditions—such as rain occurring downwind of the plant—may, however, extend this distance. The period of greatest concern is the first day or two after a release, primarily because people are unlikely to remain in the most heavily contaminated area for longer periods. Emergency planning can, under favorable circumstances, succeed in eliminating most early health effects. Unfavorable circumstances, such as a heavy snowstorm, could contribute to substantial early health effects, despite emergency planning. Poor emergency planning, however, may even enlarge risks (e.g., by directing evacuees into the path of plume, or by exacerbating traffic problems).

Delayed health effects may arise from direct exposure of people to the plume (where the time period of concern is the duration of plume passage) or from exposure to contaminated ground (where the period of concern may be many years). In each case, emergency planning can reduce the incidence of delayed health effects. These effects will not be entirely eliminated, however, even under favorable circumstances. Radioactivity, widely dispersed over the surrounding environment, may expose people to doses significantly above background levels. The same generalization applies to exposure arising from contamination of the food chain and water supplies.

Long-term protective actions may include the relocation of populations and the interdiction of food and water supplies. These long-term protective actions may help to reduce the delayed health effects but cannot eliminate them entirely. Also, these actions will exact their own adverse social, psychological, and economic impacts.

The initiating events of certain accidents may degrade emergency response. A severe nuclear plant accident may arise from an event that also reduces the capability for emergency response. For example, a severe earthquake might initiate a core-melt accident at the TMI plant. The earthquake could also destroy or seriously degrade the local capability for warning, dose assessment, and other emergency-response functions. Floods, too, are a source of significant concern and

TABLE 5

Accident Consequences and the Role of Emergency Preparedness

Short-Term Consequence	Distance of Concern	Time of Concern	Can Emergency Preparedness Help?	
Risk of Early Death	0-5 miles (0-25 miles in adverse conditions)	1-2 days	Yes	
Other Early Health Effects (including harm in reproduction)	0-25 miles	1-2 days	Yes	
Delayed Health Effects Due to Plume Exposure	100 miles +	days	Yes	
Delayed Health Effects Due to Ground Contamination	100 miles +	hours-years	Yes	(emergency evacuation may be a prelude to long-term population relocation)
Delayed Health Effects Due to Contamination of Food Chain and Water	100 miles +	hours-years	Yes	(emergency food-chain interdiction may be a prelude to long-term interdiction)

Long-Term Consequence	Distance of Concern	Time of Concern	Can Emergency Preparedness Help?
Long-Term Population Exposure from Ground, Food-Chain and Water Contamination	100 miles +	months-years	Yes (exposure may be reduced through population relocation of food chain interdiction)
Social and Economic Dislocation Due to Long-Term Population Relocation or Food-Chain Interdiction	100 miles +	months-years	No

uncertainty. Terrorist or military attacks could have similarly disruptive effects. In such cases, only small benefits of emergency planning may be achieved.

THE ORGANIZATION OF EMERGENCY PLANNING AND RESPONSE

8. Who has what responsibilities in emergency planning?*

In the event of an accident, five principal groups have responsibilities for emergency response:

- the utility is responsible for on-site safety, bringing the plant under control, and informing the off-site authorities;
- the federal agencies give technical support and assistance to the utility and the state agencies;
- the off-site authorities (including the Governor, state agencies, and state and local government), independently from the utility, continuously monitor plant status in order to identify accident conditions; thereafter they monitor on-site and off-site aspects of the accident. Taking account of the condition of emergency preparedness, these authorities recommend appropriate protective actions to be taken by the public and inform the media;
- the media are responsible for informing themselves and the public about accident conditions and appropriate protective actions; and,
- members of the public are responsible for informing themselves and taking appropriate protective actions.

The specific groups and their respective responsibilities will vary from site to site and from state to state.

---

*As indicated in the introduction, this section and all other sections in this model plan assume that the appropriate roles, responsibilities, and resources have been allocated as necessary. Reference is made to these roles, responsibilities, and resources, as if they are established, even if this is not currently the case under existing plans.

A nuclear accident in Pennsylvania, for example, will call into a play an array of on-site and off-site activities as various groups exercise their primary roles and responsibilities (Figure 6). Among the groups are:

The utility. During an accident, the primary responsibility of the utility is to bring the plant under control and to minimize the risk of a radioactive release that might threaten the public. Since plant personnel have first-hand knowledge of plant procedures and existing plant conditions, they are responsible for providing public officials with comprehensive and accurate information on a timely basis. This involves an obligation to provide an initial classification of the severity of the event. The utility is also responsible for protecting plant personnel and others at the plant site.

The Pennsylvania Emergency Management Agency (PEMA). At the Three Mile Island plant, PEMA is the lead agency for off-site planning and response. PEMA oversees the preparation of emergency plans to ensure their effectiveness and coordination. In the event of an accident, PEMA coordinates the multitude of agencies to achieve an effective emergency response. PEMA acts as a clearinghouse for information, maintains the communications network, tracks emergency resource needs, and generally acts as an intermediary between the utility, the public authorities, and the public.

The Bureau of Radiation Protection (BRP). In Pennsylvania, BRP continuously monitors nuclear plants in order to identify accident conditions; thereafter, it conducts a technical assessment and evaluation of the accident based on information received from the plant and off-site monitoring. BRP assesses the likely severity, timing, and duration of an accident, and the projected doses and distribution of radiation; this may lead to a classification of the event as more severe than that proposed by the utility. According to this assessment, BRP evaluates the alternative protective-action strategies, makes recommendations to PEMA, and advises the Governor. PEMA evaluates these recommendations against other considerations, such as the time of day, the prevailing weather conditions, and so forth, to determine what protective actions are feasible and most appropriate. Recommendations from BRP and PEMA are conveyed to the Governor.

The Governor. The Governor authorizes the state agencies to recommend and implement the appropriate protective actions. The Governor has the authority to declare a state of emergency and to call the National Guard to active duty.

The federal agencies. Federal agencies, such as the Department of Energy (DOE), the Environmental Protection Agency (EPA), and the

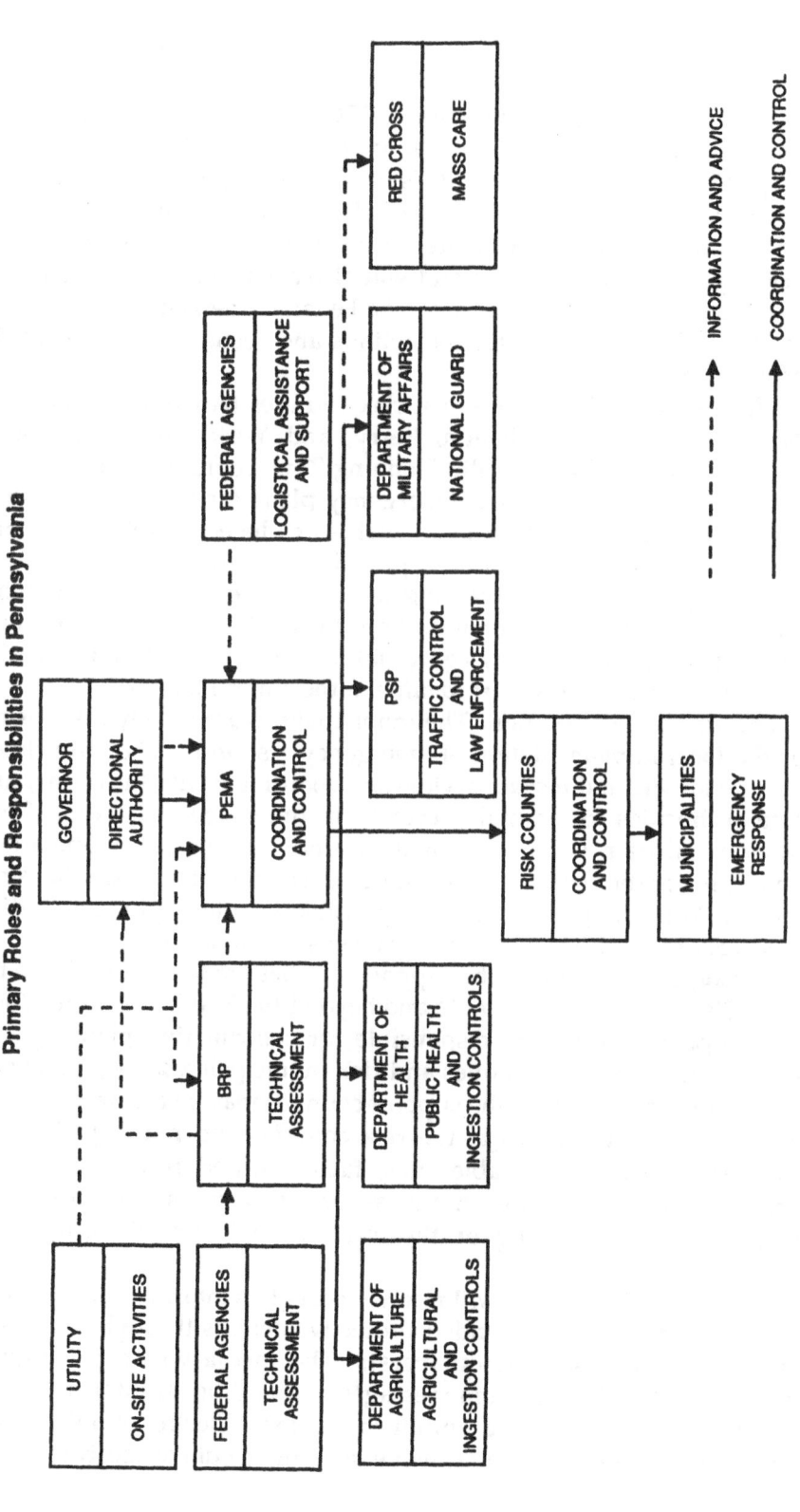

FIGURE 6

Primary Roles and Responsibilities in Pennsylvania

Nuclear Regulatory Commission (NRC), and national laboratories, such as Brookhaven National Laboratory, may assist the utility and the public authorities. The NRC is ordinarily responsible for the regulation and licensing of nuclear power plants. In the event of an accident, the Commission will coordinate the activities of the federal agencies assisting the on-site response and will also act as liaison between these agencies and those acting off site. The Federal Emergency Management Agency (FEMA) coordinates the actions and recommendations of the federal agencies off site.

The risk counties. The seven risk counties (Adams, Cumberland, Dauphin, Lancaster, Lebanon, Perry, and York) within the Inner Planning Zone and the Middle Planning Zone coordinate the development and implementation of emergency plans for the municipalities, school districts, and institutions (such as colleges, hospitals, nursing homes, and prisons) within their jurisdictions. In general, they assist the *municipalities* in implementing the protective action recommendations by maintaining a communications network, by matching resource needs and supplies, and by coordinating activities with adjacent risk counties and with the *support counties* in the Outer Planning Zone.

The risk municipalities. The municipalities are largely responsible for the local implementation of emergency response. The municipalities provide fire and rescue services and coordinate with the county and state authorities to maintain security in the evacuated areas. They provide and train volunteers for the municipal emergency-operations centers, for traffic control, and for route alerting to supplement the siren system. Municipalities are also responsible for identifying and aiding those residents with special needs, such as the handicapped.

The support counties. The support counties are those counties in the Outer Planning Zone adjacent to and beyond the Middle Planning Zone. The support counties are responsible for coordinating planning and response with the risk counties and risk municipalities. Depending on the nature of the accident, the support counties may be required: to provide additional resources to the risk counties and municipalities, as necessary; to make available sites, facilities, and resources for the screening, decontamination, and mass care of evacuees; and, to assist in post-accident monitoring of the environment and of agricultural produce.

The media. The public will receive most of its information about any accident and any recommended protective actions through the media (television, radio, and newspapers). The media will carry regular Emergency Broadcast System messages and additional information on conditions at the plant and within the region. Electronic links to the Emergency Operations Center, will enable the media to receive regular updates on the plant, weather, off-site radiation, and traffic conditions. This information will be relayed to the public in a timely manner and will be continuously updated.

The public. The goal of emergency planning and preparedness is to protect the public. Members of the public, however, also have their own responsibilities. They should seek and read the information distributed by the state and the utility and retain copies of this information for future reference. Individuals should also develop their own personal and family emergency plans, including potential routes and destinations to be used during an evacuation and ways to keep informed about a developing accident (see sections 18-23 below).

Other organizations. The state Departments of Agriculture, Health, and Military Affairs, the Pennsylvania State Police (PSP), and the Red Cross also have primary responsibilities in the event of an accident. Other state and local government agencies have secondary responsibilities. The details about these organizations and their responsibilities for any particular nuclear plant would be provided in the implementing procedures.

---

9. What are the Emergency Planning Zones?

The map in Figure 7 shows the locations of the three Emergency Planning Zones for the Three Mile Island region:

- The Inner Planning Zone is the area within five miles of the plant. It is the area of highest risk, which requires the most rapid response. The primary emergency preparation, whenever feasible, is for early evacuation of the entire zone, before core melt begins.
- The Middle Planning Zone is the area located between five and twenty-five miles from the plant. This is an area of lower risk requiring flexible response. The primary emergency preparations are for sheltering and evacuation downwind from the plant.
- The Outer Planning Zone is the area beyond twenty-five miles from the plant. The primary emergency preparations are for sheltering, followed by evacuation or relocation from hot spots, and protection from radioactivity in food and water.

44

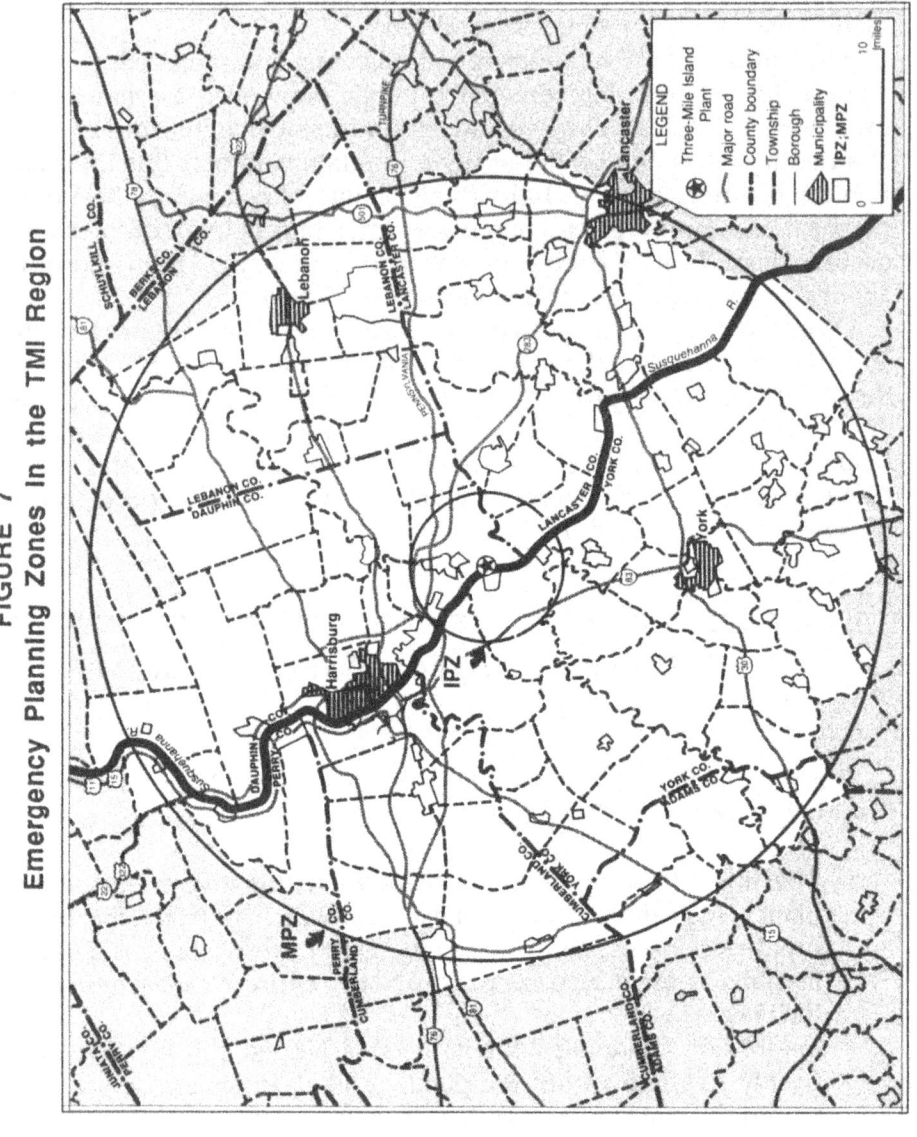

FIGURE 7
Emergency Planning Zones In the TMI Region

Purposes of emergency-planning zones. Emergency-planning zones serve two primary functions:

- they help people to know how their location affects what they should do and what hazards concern them in the event of a nuclear emergency; and,
- they also provide the geographical organization for emergency-response activities.

The definition of the zones is based on an assessment of the most serious threats to human health and the best ways of avoiding such threats or reducing their impact. This plan defines three zones, reflecting the different types and intensities of potential radiation exposure at different distances (see Section 5) and the potential health effects of radiation at different dose levels (see Section 6).

The Inner Planning Zone. The area nearest the plant will usually incur the most serious radiation exposures. Even for the most severe releases, exposures that could cause early death from radiation are likely to be confined within this zone. The most effective emergency response within this region is for people to evacuate quickly. Much planning effort is directed to anticipate conditions that may precede a severe accident and to order evacuation before a release occurs. If these efforts fail, so that a major release of radioactivity is actually under way, or if obstacles are preventing evacuation, then sheltering followed by evacuation may be more appropriate. Emergency preparedness and resources need to be most highly developed within this zone.

Emergency preparations within the Inner Planning Zone include the development of a strong capability to:

- notify the public and alert emergency workers immediately;
- mobilize rapidly all the resources necessary for timely evacuation;
- mobilize screening, decontamination, and mass-care facilities;
- educate and inform the public and train emergency workers; and
- practice these emergency-response capabilities in frequent, realistic drills and exercises.

The public within the zone must be prepared to:

- respond to notification and keep themselves informed of unfolding events;

- shelter or evacuate promptly when directed to do so (choosing a route, as advised, that avoids the expected plume movement); and,
- use potassium iodide and respiratory protection when so advised.

The Middle Planning Zone. Beyond the inner zone, exposures leading to early death are unlikely, unless a person remains in an area of high radiation for many hours, but the uncertainties about where and when the radioactive plume will travel are greater. The basic approach is to monitor the developing accident and then to respond quickly in a broad downwind area. The most effective strategy for limiting radiation exposures is usually to shelter and then evacuate. A severe accident or adverse weather conditions might well require an early evacuation.

Emergency preparations within the Middle Planning Zone are similar to those for the Inner Planning Zone except that there will probably be more time available to respond. Such preparations include the development of the capability to:

- notify the public and alert emergency workers immediately;
- mobilize rapidly all the resources necessary for timely evacuation;
- monitor environmental radiation levels and track plume movement closely;
- mobilize screening, decontamination, and mass-care facilities;
- educate and inform the public and train emergency workers; and,
- practice these procedures in frequent, realistic drills and exercises.

The public should be prepared to:

- respond to notification and keep themselves informed of unfolding events;
- shelter or evacuate promptly when directed to do so (choosing a route, as advised, that avoids the expected plume movement); and,
- use potassium iodide and respiratory protection when so advised.

The Outer Planning Zone. Beyond the two inner zones, the primary concern is with radioactive material deposited on the ground. Much of the exposure from this material is likely to come through contaminated food. Monitoring of agricultural produce, therefore, will be important. Specific areas or "hot spots"of high radioactivity will need to be identified, and evacuation of such areas may be necessary. Short-term or long-term relocation of certain populations may be required in this zone (see Section 25).

Emergency preparations for this zone include:

- public notification through the media;
- a capability for plume tracking and for monitoring radioactivity on the ground and in food supplies and water, and;
- preparation by farmers to protect their animals and crops.

The public should be prepared to:

- follow recommendations for sheltering, evacuation, and the use of potassium iodide;
- follow directions for avoiding exposures through food and water; and,
- relocate from contaminated areas, if necessary.

---

10. What special considerations exist for emergency planning in the TMI region?

The region around TMI presents several particular issues for emergency planning and response. The region is:

- the only place in the United States in which a nuclear power plant accident has already triggered an evacuation;
- geographically complex, with many hills and narrow valleys that complicate local meteorology and could impede an evacuation;
- agriculturally rich and produces many food products that would require careful protection and monitoring in the event of an accident;
- home to a large and diverse population, including several cities and urban areas that might need to be evacuated in the event of an accident; and,
- has several special populations that would require particular attention in an emergency, including not only school children, hospital patients, nursing home residents, and others found in all regions, but also a distinctive cultural group—the Amish.

The United States has some 111 operating nuclear power plants located at more than 65 different sites. Each site has unique features of geography, meteorology, population distribution, and social characteristics that complicate emergency planning and response. In this regard, the TMI region is neither the most, nor the least, difficult region for which to plan.

Emergency plans cannot be engineered. They must grow out of, and respond to, the fabric of land, people, transport systems, cities and villages, and cultures that exist in the region, of which the nuclear plant is only a small and recent part.

The legacy of the 1979 accident. Although the TMI region may not be the most difficult region for which to plan, it is the one site in the United States in which a serious nuclear power plant accident prompted a governmental advisory that resulted in a partial public evacuation. Indeed, subsequent analyses by the US Nuclear Regulatory Commission indicated than an evacuation of populations near the TMI plant should have been ordered. This experience at TMI has affected the perceptions of and anxiety about nuclear power and nuclear accidents among many members of the public. The accident has left some people skeptical and distrustful of the utility and the government. This legacy creates some special uncertainties in the TMI region as to how different population groups would respond to another accident.

On the other hand, the public in the region is probably better informed and better prepared for an emergency than the public near many other nuclear plants in the United States. The utility, state and local governments, and numerous volunteers have devoted substantial time and energy to improve the region's emergency-response capability.

A geographically complex region. Located in a region of gently rolling hills in south-central Pennsylvania, the plant sits astride Three Mile Island, in the Susquehanna River. Its cooling towers dominate the river vista and have become a favorite haunt for tourists. The Susquehanna bestows a bounty of plentiful water for cooling the TMI plant, but the river also extracts its price. Flooding may initiate an accident and at the same time impede evacuation, should it be necessary. Even in the absence of flooding, the river constrains evacuation, since there is only one bridge within five miles of the plant.

The region is also meteorologically complex. However picturesque the Susquehanna valley and the nearby hills, predicting the likely path of a radioactive plume becomes difficult. A particular problem may be the Susquehanna's channelling effect where some adjacent hills rise to over 500 feet on both sides of the river. Channelling northwards might bring the plume into the proximity of several densely settled

urban areas. In addition, hurricanes, snowstorms, and icestorms periodically traverse this part of Pennsylvania. These major storms can cause
a loss of off-site power to the plant and can also complicate the task of
evacuating the population.

Agriculture. The historic Susquehanna Basin was the bread basket
of colonial America. It remains one of the richest agricultural areas in
the nation, with large numbers of dairy farms, chicken farms, and or-
chards. Above 500 feet, croplands and pasture give way to forests, and
to the south pastures compete with patches of woodlands (Figure 8).

The agricultural heritage of the region also poses potential problems
in a nuclear emergency. To avoid contaminating milk, farmers may be
called upon to remove cows from pasture and to feed them from stored
stocks. Milk, such as that used at the Hershey chocolate factory, will
need to be screened. Other agricultural products, such as fruits and vegetables sold at hundreds of farm stands, will need to be screened and
destroyed if found to be contaminated. Of course, the particular type of
protective actions necessary would also depend on the season. If an
accident happened in winter, cows would already be inside and on
stored feed, and farm stands would be closed.

Demography. In 1990, more than 35,000 people lived in the three
counties (Dauphin, York, and Lancaster) and the 13 municipalities and
townships within 5 miles of the plant (Figure 9). Meanwhile, 1,428,656
people live in the counties (Dauphin, Lebanon, Lancaster, York, Adams,
Cumberland, and Perry) and municipalities within 25 miles of the
plant. Of special concern to emergency planners is the ribbon of dense
settlement stretching from Middletown (population 9,254) 10 miles
north along the eastern shore of the Susquehanna to Harrisburg
(population 52,376).

For all this, the TMI region is not the emergency-planning nightmare
of more populous nuclear plant sites in Europe and the United States,
although the region lies in close proximity to some major metropolitan
areas (Figure 10). Nonetheless, this is a sizeable population requiring
protection. The region includes several groups with special needs, including institutional populations (such as in schools, colleges, hospitals, nursing homes, and prisons), farmers, the Amish, and tourists.

The Amish. The Amish are a religious community with distinct beliefs and customs. They favor more simple, traditional ways of rural
life. More than 20,000 Amish live in Pennsylvania—including 15,000 in
the rural areas of Lancaster County, between 20 and 30 miles southeast
of the plant. Providing adequate protection for these Amish communities, in the event of an accident, is a particular challenge for emergency
planners:

FIGURE 8

Local Political Boundaries and Transportation Routes

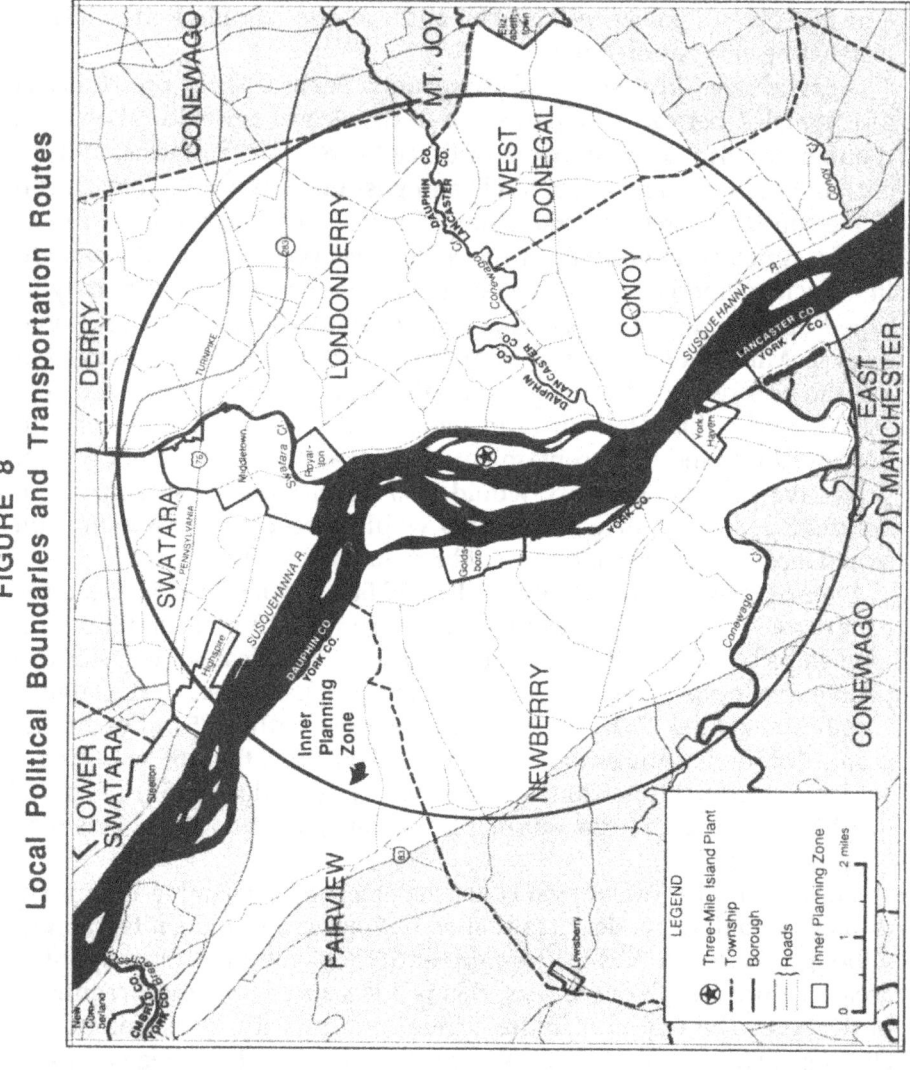

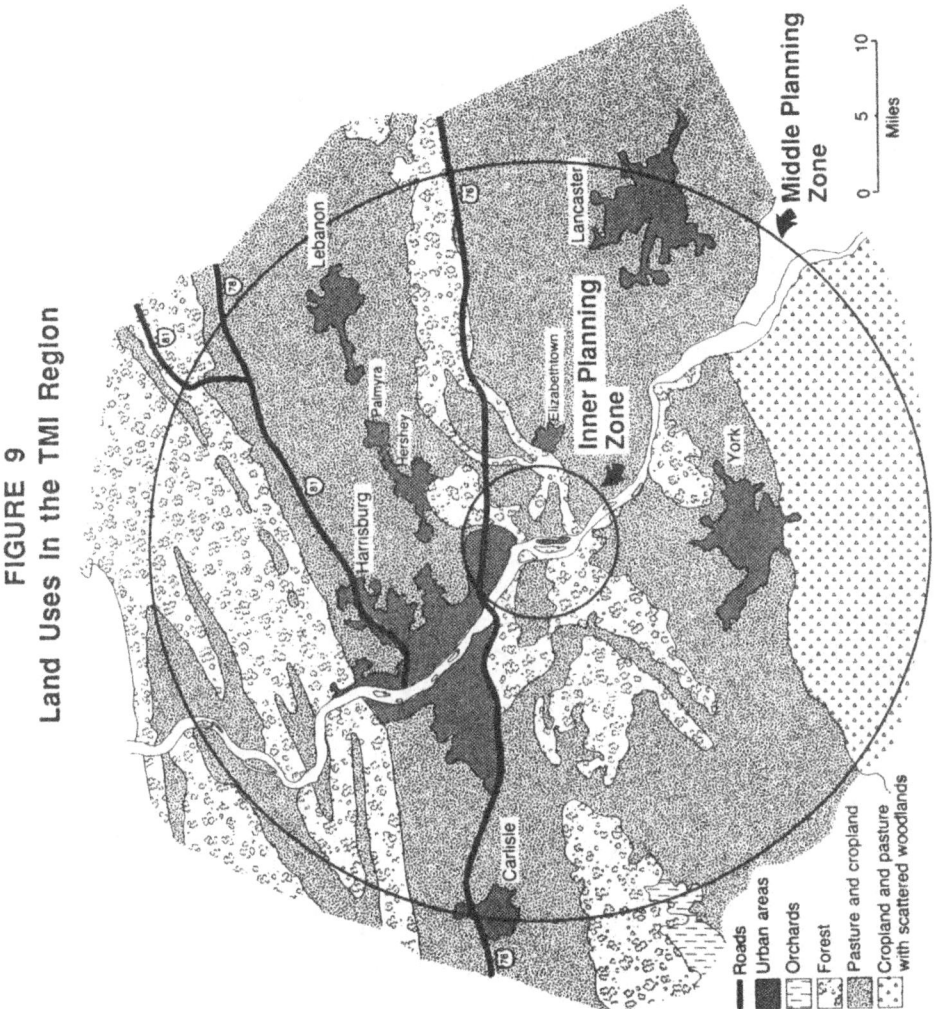

FIGURE 9
Land Uses In the TMI Region

FIGURE 10
The TMI Region and Neighboring States

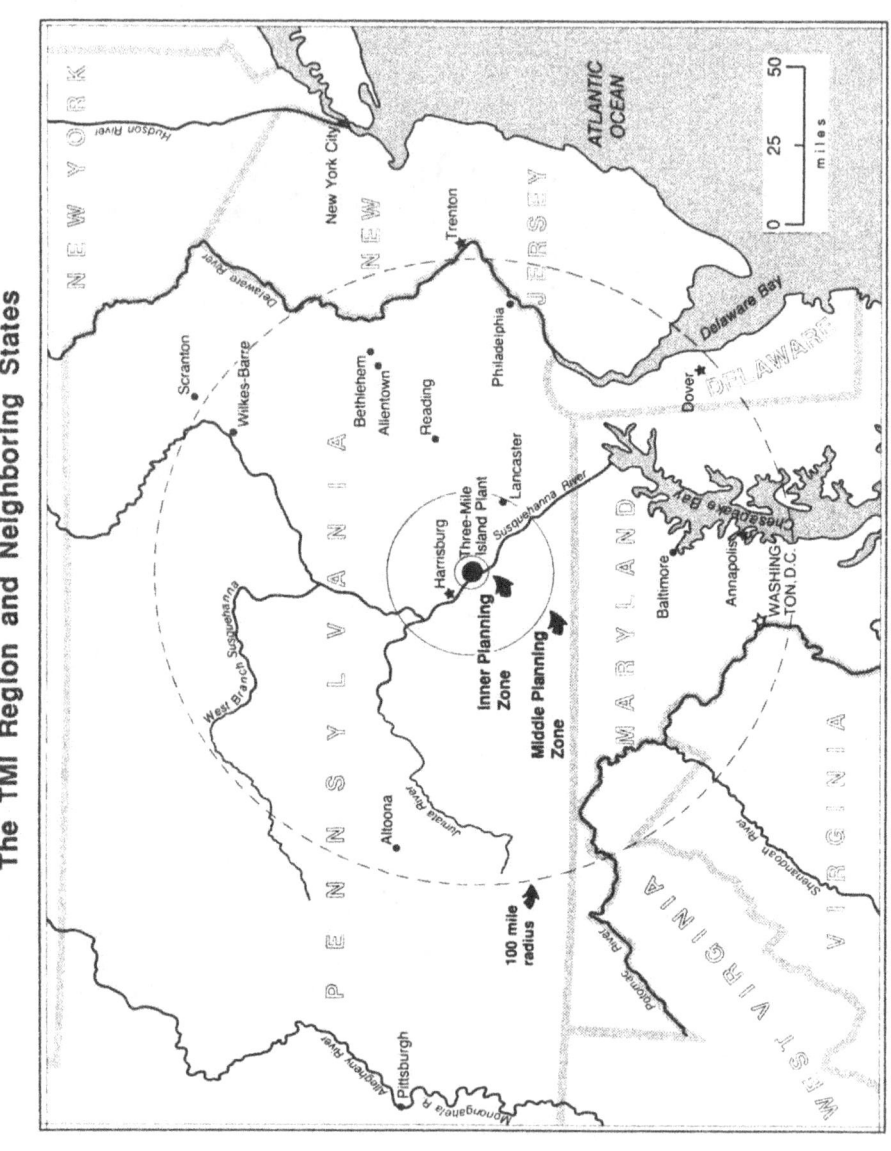

- many of the Amish are farmers who require additional information on the protection of animals and agricultural products;
- the Amish reject the use of television, radios, and automobiles, making notification of an alert or evacuation more difficult; and,
- because Amish customs and beliefs are so distinctive, appropriate educational and information materials will be required.

Other sections of the plan suggest ways to meet these challenges.

ACCIDENT EVALUATION

11. How will the safety status of the plant be monitored and evaluated?

Emergency preparedness requires an effective system by which responsible public authorities can monitor the safety status of a nuclear plant, anticipate severe accidents, and independently validate the threat assessments of the utility. This will involve:

- the stationing of a technical representative of the state radiation agency—in Pennsylvania, the Bureau of Radiation Protection (BRP)—at the nuclear plant;
- a computer link between the plant and the state agency so that a continuous flow of critical plant parameters will be available; and,
- the development of appropriate analytical and personnel resources to assure that the state agency has the capability to monitor and to evaluate systematically the monitoring data.

The role of monitoring. The course of a severe accident may not at first appear to be heading towards a release. In fact, many situations will occur in which an adverse event in a plant is controlled and a release avoided. A plant monitoring system is needed to identify adverse trends and events and to provide information to aid state officials in evaluating the seriousness of a specific incident. The plant monitoring system can also provide valuable ongoing information about the development of an accident.

As Section 4 indicates, an accident might develop quite quickly. A release could occur in less than an hour after the initiating event. The plant monitoring system must therefore be able to provide prompt warning of any adverse trends. Yet the operation of a nuclear plant is characterized by long periods that are uneventful. Thus, the monitoring system must be capable of responding reliably after long periods of routine activity.

Emergency responses, especially evacuation, are more effective if they can go forth in advance of a release. This plan thus allows for the precautionary initiation of emergency responses, even though an actual release may not always materialize. To this end, an event will be classified as a Projected General Emergency if the probability of the event proceeding to the core melt is estimated to exceed 10 percent (see Section 13). An important function of the plant monitoring system is to provide information whereby such an estimate can be made promptly.

The utility's technical staff are, of course, trained to detect adverse trends in plant operation, to take corrective actions, and to evaluate the development of an emergency. Extensive instrumentation is available to assist them. It is essential, however, that state public authorities have independent information about plant status. This provides added confidence that any adverse trend will be detected early and carries the further advantage that public authorities can reach their decisions free of the potential stresses or conflicts of interest that might confront utility staff. This calls for a type of plant monitoring system similar to that currently in operation at the Illinois Department of Nuclear Safety (see Box).

Plant monitoring capability. Under this plan, the responsible state agency, such as Pennsylvania's Bureau of Radiation Protection, or BRP, will be provided with continuous on-line access to hundreds of important plant parameters. The plant computer will feed these parameters to a computer in the responsible state agency. Any adverse

---

THE ILLINOIS SYSTEM
FOR MONITORING NUCLEAR PLANTS

The Illinois Department of Nuclear Safety has developed a system to monitor continuously the safety status of the thirteen commercial reactors in Illinois.

Via electronic data links, each reactor's computer transmits to the Department headquarters in Springfield more than a thousand reactor parameters. Each parameter is transmitted once every two to four minutes. The availability of the data stream from each plant has typically exceeded 95 percent.

At present, Department personnel monitor the data streams manually. Computer software under development will eventually provide automatic indication of adverse plant trends as well as real-time projections of the future development of such trends.

trend will be automatically detected, and this occurrence will be communicated to an on-call expert in nuclear engineering. This individual, using a computer terminal in the office or at home, will evaluate the situation and determine the necessity for further action.

Once it has been determined that an emergency is under way, a more elaborate system of evaluation will come into action. One or more state officials will travel to the plant site and to the utility's off-site command center. From this center, they will communicate with utility officials and with their colleagues in the state capital. The state will continue to receive a flow of plant parameters and will also have additional information from utility sources.

Evaluation of plant monitoring data. The evolving status of the plant will determine how the emergency is classified (see Section 13 below). Consultation among utility, state, and (possibly) federal officials will precede this decision, which will rely on computer-based analysis. The utility has the responsibility to make an initial classification of the accident. State officials have the authority either to concur with this judgment or to substitute a higher (or lower) classification.

If a release does occur, the plant monitoring system will provide information that can be used to help estimate the magnitude and composition of the release. This information, together with the results of off-site environmental monitoring, will assist public officials in estimating off-site protection needs.

12. How will the plume be tracked and dose projections be made?

Once an atmospheric release has occurred, rapid assessment of the potential public health impact is essential. The utility and state authorities will need to provide for:

- rapid detection of a release, using fixed monitors;
- rapid tracking of the plume, using quick-reaction aircraft and mobile ground teams;
- immediate transmission and processing of monitoring data; and,
- prompt projection of the geographical pattern and timing of radiation exposures, from both the passing plume and deposited material.

Initial detection of the release. Deteriorating plant conditions will sometimes signal that a release may be imminent. In some instances, plant instruments will also indicate the commencement of a release. An independent system to detect the onset of a release is, however, an essential part of a well-designed monitoring system. For this purpose, fixed radiation monitors are located in a two-mile radius around the plant. Data from these monitors will be continuously relayed to the utility's on-site and off-site control centers and to the state regulators offices.

Plume tracking and dose projection. First it is necessary to confirm that a release is under way. Then the task is to measure and predict, through an integrated program of field monitoring and computer modelling, the dispersal of radioactivity in the environment and the potential exposure of nearby populations. Computer models will be used to predict the behavior of the radioactive plume (see Box). These predictions will undergo progressive refinement according to the field results obtained from the environmental monitoring effort.

An important element of the monitoring effort will be a *quick-reaction airborne monitoring capability.* Under this plan both the utility and state officials will have instrument packages and trained staff always available for use with any standard helicopter or light

---

THE NEED FOR NEW DOSE-PROJECTION MODELS

Although improved dose-projection models are being developed, existing models fall well short of the needs of decision makers, emergency workers, and the public. In particular, existing models are deficient because they rely on dose projections based on inadequate monitoring and meteorological information. New models will incorporate the extensive measuring capability called for in this plan. Other major improvements needed include:

- graphical displays so that projected areas of radiation exposure can be readily televised to the public
- the rapid display of expected uncertainties in dose projections
- the rapid revision of dose projections to incorporate field-monitoring data
- the integration of short-distance and long-distance modelling capabilities.

aircraft. Contractual arrangements will ensure that an aircraft can always be supplied on short notice. After several hours of use, this quick-reaction airborne monitoring capability will be supplemented by special-purpose federal government aircraft from a designated Air Force base. By the second day of an accident, it is anticipated that the airborne monitoring effort would be almost entirely in the hands of the federal government.

Both the utility and state authorities will be responsible for deploying *ground-mobile monitoring teams.* Each ground-mobile team, and the airborne monitoring teams, will be equipped with up-to-date telemetry devices for transmission of data to command centers, which process data by computers, to avoid the delays and errors associated with manual processing. Eventually, these teams will be supplemented by teams from other states and utilities, and by federal teams.

Modelling by both the utility and the state agency will undergo continuous updating based on monitoring data. The agency will also be linked electronically to the computer modelling capability at Lawrence Livermore Laboratory in California, which can be activated on short notice on a round-the-clock basis. The results of modelling will take the form of maps showing contour lines of potential radiation exposure. These maps will be in a format directly usable by TV stations and will be part of the ongoing information-dissemination program.

---

13. How are nuclear accidents classified?

The choice of protective response measures depends on many considerations, such as the status of the plant, the prevailing weather, the time of day, and the state of traffic. This plan recognizes five major classes of accidents. Four classes refer to the potential or actual occurrence of the most serious accidents, those involving core melt. A fifth refers only to accidents in progress, without core melt.

- unusual event: an event that might compromise plant safety
- alert: a significant increase in the probability of a core melt or a radioactive release
- projected general emergency: a substantial increase in the probability of a core melt
- general emergency: core melt imminent or ongoing

- limited-area emergency: release of radioactivity imminent or ongoing, but with no core melt

Roles and responsibilities.   The utility has the responsibility to classify conditions at the power plant relative to core melt and to estimate the probability of a radioactive release.   It also must notify relevant state agencies (in Pennsylvania, BRP and PEMA) of the classification.   Responsible state officials may, on the basis of their own independent assessment of conditions at the plant, upgrade the level of classification.   At the Alert level or as soon as possible thereafter, state officials are responsible for assessing weather conditions and for scanning the potentially affected region to identify any particular problems and to assess the state of readiness.

Purpose of the classification.   Accident classification provides everyone involved in emergency response with a quick picture of the current status of the accident.   It also provides decision makers a ready structure for guiding their choice of protective actions.

The appropriate protective response depends on two sets of conditions:

- conditions at the plant that determine the potential for a serious accident; and,
- conditions external to the plant—weather, road conditions, and the state of emergency resources that determine the ability of people affected to mount appropriate protective actions.

The most serious accidents are those involving partial or full core melt.   Accordingly, the classification of plant conditions is largely based on plant status relative to core melt (see Table 6).

In regard to off-site conditions, there are two major concerns:

- Do meteorological conditions suggest that pre-planned protective responses should be altered?
- Are any special conditions in the region likely to impede timely protective responses?

Expected sequence of classification.   The order of the four classes referring to core melt in Table 6 is based on increasing severity of conditions.   If an accident proceeds as anticipated, it may be possible to move from class to class as the accident develops.   But accidents often do not proceed as anticipated.   Some events may happen so rapidly that a General Emergency could occur with very little warning.   Many accidents however, can be controlled and do not proceed to more severe

TABLE 6

Classification of Nuclear Accidents

Classification Level	Plant Conditions	Level of Response
Unusual Event	an event that might compromise plant safety	notification of relevant state agencies and the US Nuclear Regulatory Commission
Alert	1) significant increase in probability of core melt or 2) significant possibility of radioactive release without core melt	activation of emergency-response organizations
Projected General Emergency	substantial increase in probability (10% or greater) of core melt	initiation of precautionary emergency response
General Emergency	core melt imminent or ongoing	appropriate response in all zones
ADDITIONAL CLASSIFICATION FOR LESS SEVERE ACCIDENTS (NO CORE MELT)		
Limited-Area Emergency	release of radioactivity imminent or ongoing, but no core melt	appropriate emergency response (in inner zone)

conditions. The definition of Projected General Emergency is such that protective responses including evacuation may well be initiated without a radioactive release's subsequently occurring.

14. How will a decision to initiate protective response be made?

Decision making during a nuclear reactor emergency will typically involve high levels of uncertainty, rapidly changing conditions, and conflicting information and advice. To promote effective and timely decision making under these adverse conditions, this plan clearly identifies:

- the primary roles and responsibilities for emergency-planning and -preparedness organizations;
- the key considerations in the decision problem;
- the choice of pre-planned emergency responses based on rapid assessment and accident classification;
- subsequent emergency measures that can be implemented with greater flexibility based on extensive monitoring of a developing accident; and,
- special measures that may be required for unexpected accident conditions on and off site.

Roles and responsibilities. Many individuals and organizations will take part in decision making during a nuclear reactor accident. Since time will often be short, it is essential that these roles mesh as well as possible and that each organization have implementing procedures that address the most important contingencies. Section 8 outlines these roles and responsibilities for Pennsylvania in detail, but to reiterate:

- the utility and state radiation officials will assess the threat, the utility will classify the accident, and state officials will accept or alter the classification;
- state radiation officials will forward to the state emergency-management agency (PEMA in Pennsylvania) the classification of the accident, an assessment of weather conditions and potential radiation exposures, and any associated protective-action recommendation;

- state emergency-management officials, in close consultation with state radiation experts, will recommend a protective-action strategy designating actions to be taken, areas to be covered, timing, and special population needs to be met; and,
- the recommendation will be transmitted to the Governor who will accept or alter the recommendation.

As the accident unfolds, state officials (PEMA and BRP) and the Governor should seek to develop decision-making resources. Additional expertise may be drawn in from universities, the national laboratories, federal agencies, and utility groups, with whom prior arrangements will need to have been made. It is important to recognize that the decision process will need to be iterative, with constant reformulation of protective-action strategies in light of new information, changed weather conditions, or implementation problems.

The decision problem. Whenever a major release, resulting from a core melt, is a possibility, the most urgent decision is whether or not to recommend an evacuation and for what regions. The responsibility for this decision rests with the Governor. Because this decision may have to be made quickly and most likely will have to rely on complex and incomplete technical information, the Governor will need good advice from the plant operators and responsible state officials and they in turn must be prepared to make rapid assessments of the emergency situation. Because a precautionary evacuation, before a release occurs, is the best protective response, they will have to make assessments at a time when confusion may prevail about what is actually happening.

There are three major early decisions:

- to determine the need for an evacuation;
- to determine whether an immediate evacuation can be conducted safely, because evacuation can in some circumstances increase rather than decrease exposures;
- to determine the regions to be evacuated or which will require other protective responses.

Much of the planning focusses specifically on these decisions. Accident classification will identify the need for evacuation and/or other responses. Emergency planning zones define where protective actions are expected. State officials in their preparation and assessments must *identify and evaluate special conditions* that might alter the protective-action recommendations envisioned in the classification and the delineation of emergency-planning zones.

In the later stages of an accident the problems posed can be known with much greater certainty, and a much wider range of responses can be elicited. This plan calls for a highly developed accident-assessment capability that will provide rapid indications of amounts of radioactivity released and where it is going. Further emergency measures, such as later evacuations, sheltering, relocation of populations, and food-chain interventions will be tailored in a flexible manner to the actual accident conditions, as they become known.

Anticipated responses. Many emergency responses can be pre-planned based on the theoretically anticipated sequence of events during an accident. This anticipated sequence is reflected in the accident classification (see Section 13), which drives decision making, and which provides the organizational structure for state agency procedures.

Early evacuation, if it can be implemented, is the best protective action. The accident classification scheme focuses on identifying the need for evacuation as early as possible. Accordingly, plant operators and state officials will monitor plant conditions continuously. Any significant reduction in plant safety, which may come from a failure of equipment, or from external conditions such as loss of off-site power, will lead to a declaration of an *Alert*. This declaration calls for an activation of emergency-response organizations.

Further monitoring of plant conditions will track, to the extent feasible, pre-identified potential accident sequences. Given a compromise at plant safety sufficient to threaten a 10% or greater chance of core melt's beginning in the near future, the plant operators and state officials will raise the classification level to *Projected Emergency*. At this time, state officials should recommend an evacuation, unless adverse meteorological or other off-site conditions threaten to impede the evacuation significantly.

Plant operators and the state officials will continue to monitor plant conditions and, if core melt becomes imminent, declare a *General Emergency*. Based on plant conditions, any releases of radioactivity that occur, changes in weather conditions, and the effectiveness of the initial response, the state officials will formulate further emergency-response recommendations and organize their implementation. A major release is likely to necessitate a prompt evacuation downwind for at least a portion of the middle zone. Subsequent identification of "hot spots" will trigger still further evacuation. If, on the other hand, the danger of a core melt is avoided, then notification of the end of the accident can give way to a winding-down of response activities.

What special conditions can change this sequence? What will happen in a nuclear emergency is not very predictable, especially in the early stages. In their implementing procedures, state planners will have prepared a set of initial responses to cope with developments that do not follow the anticipated sequence. Here are three particularly important examples:

- **The accident may develop more rapidly than expected or may occur without sufficient warning.** Plant operators may suddenly find themselves in a situation in which a core melt is imminent or under way, or even one in which a release has begun. If core melt has been under way for some time or if a release is already in progress, assessors may decide that it is safer to shelter people immediately and plan to evacuate them after the bulk of the radioactivity has passed.
- **There may be impediments to evacuation.** Serious delays in evacuation for a particular area, or a particular group of people, may result from bad weather, flooding, or transport blockages. State officials may recommend that a subpopulation shelter during the early stages of the accident while provisions are made to expedite a later evacuation.
- **There may be adverse weather conditions.** State officials may determine a substantial possibility that the plume of radioactive material will be narrow and of high concentration, or that rain might bring down radioactivity. In such circumstances, they would recommend that the precautionary evacuation be extended to a down-wind area in the middle planning zone.

THE EMERGENCY INFORMATION SYSTEM

15. How will the emergency information system assure a continuing flow of accurate information about the changing accident situation?

The emergency information system will comprise:

- a centralized computer system at the state agency (PEMA in Pennsylvania), with backup computerized and manual systems, to gather, evaluate, and disseminate information about on-site and off-site conditions;
- multiple lines of communications between parties and backup personnel to provide redundancy in case of failures;
- multiple horizontal and vertical communications links to provide appropriate verification and flexibility; and,
- a computerized graphics capability to map rapidly information sent and received.

The emergency information system will track:

- on-site conditions at the plant
- off-site conditions, including:
  - local weather;
  - plume size, shape, direction, and location;
  - traffic and other road conditions; and,
  - the progress and problems of local response.

This information will be relayed continuously to:

- on-site emergency workers and decision-makers;
- off-site emergency workers and decision-makers; and,
- members of the public.

Goals. The underlying philosophy of this model plan is that most people—emergency workers, decision makers, and members of the public—will usually respond reasonably and effectively if they receive good information and if they understand why particular actions are necessary. The ability of people to respond quickly and effectively in an emergency is greatly improved when they have a comprehensive picture of the emergency situation. In particular, they need to see how their individual activities and those of others fit together and how particular actions will provide them with greater protection. In this respect, a rich, dynamic, and controlled information environment is perhaps the most important resource in emergency response.

The goals of the emergency information system, therefore, are to:

- promote understanding of the developing accident and available protective actions;
- provide information about problems and opportunities that develop as the emergency unfolds; and, to
- enhance the ability of people to respond flexibly to the emergency.

To achieve these goals the emergency information system must provide accurate information throughout the duration of the emergency. Changes must be conveyed quickly and accurately to emergency workers and members of the public.

A two-way process. The emergency information system must involve a continuous, two-way flow of information. All people in the emergency response network and members of the public should have as much information as possible, including:

- the current conditions at the plant;
- the progress of the radioactive plume;
- the current conditions off site, including local weather and traffic conditions;
- the necessity for certain protective actions; and,
- the availability of emergency resources.

At the same time state emergency response agency (e.g., PEMA) will coordinate the gathering of information from:

- the utility, state agencies, and federal agencies: concerning plant status and possible accident sequences;
- utility and state environmental monitoring teams: concerning weather conditions and radiation levels in the region;

- county and municipal emergency workers: concerning traffic and road conditions, problems at mass-care centers and decontamination stations, and the public response in general; and
- institutions, such as prisons, hospitals, nursing homes, schools, and colleges: concerning their special needs and problems.

The Agency will gather, evaluate, and disseminate information about on-site and off-site conditions through a centralized computer system, with backup computer and manual systems. Instrumentation in the plant should be able to relay critical information concerning the accident even if the plant has been evacuated. Multiple redundant communication channels, including microwave transmission with a relay station, dedicated phones, fax machines, and radios, will link state, county, and municipal authorities. The amateur radio network will be integrated with the formal state system, and the operators will receive appropriate training. The authorities will maintain sufficient staff to provide two shifts and replacement personnel, as necessary. This redundancy will allow parties to communicate even if some channels fail.

Because the state emergency response agency acts as the central focus, gathering and disseminating information, it is necessary to strengthen 'horizontal' links within the hierarchy of communications to avoid over-centralization. Each county and municipality should be able to communicate with the other counties and municipalities *without* having to channel information through the state agency. These horizontal links enable the system to function even if the central facility at the state level is inoperative, and they allow flexibility to use alternative channels of communication should several channels to one party become blocked. An essential component of such horizontal communication is a pre-established and pre-distributed format for the information exchanged. Such a format serves as a checklist for avoiding errors of omission.

Finally, a computerized graphics capability is the most appropriate way to relay much of the necessary information, such as plume location, traffic problems, and the locations of screening stations. Links to local television stations will allow the rapid dissemination of information to the public.

16. How will emergency response organizations be alerted and informed about the changing situation?

In the event of an accident, the following actions will be taken:

- The utility will inform state agencies (PEMA, BRP), and the counties at risk.
- State officials (BRP) will monitor the plant status by direct computer surveillance of critical parameters.
- State officials (PEMA) will notify the counties at risk and initiate public alert and notification.
- The risk counties will notify their municipalities.
- The media will disseminate the Emergency Broadcast System messages and additional public information.

Goals and objectives. The overall goal of the organizational communications system is to assure that emergency workers will have the information needed for timely and effective actions. This information should not state only what to do, but why these actions are necessary and how they fit into the overall emergency response. Understanding why certain actions are necessary will promote flexibility in response by encouraging individuals to use initiative in unfamiliar or unexpected situations. Emphasis will be given to *horizontal* as well as *vertical* communications links to provide resiliency in information flow.

The communications network. In the event of an accident at the nuclear plant, the *utility* will inform state agencies, and the counties at risk about the nature of the accident and possible emergency actions. The plant is linked to these off-site authorities by dedicated phonelines, with backup radio communications (see Figure 11).

The state radiation agency (BRP) assigns a nuclear engineer to oversee activities at the plant and the Emergency Operations Center. In addition, this agency tracks plant status by direct computer surveillance and monitors environmental radiation levels off site (see Sections 11 and 12). This system provides early independent warning of potential problems at the plant, and aids in the clarification and verification of information received from the utility. Based on this information, state radiation officials will conduct a technical assessment of the situation and make recommendations to the state emergency-management agency (PEMA).

FIGURE 11

The Notification System*

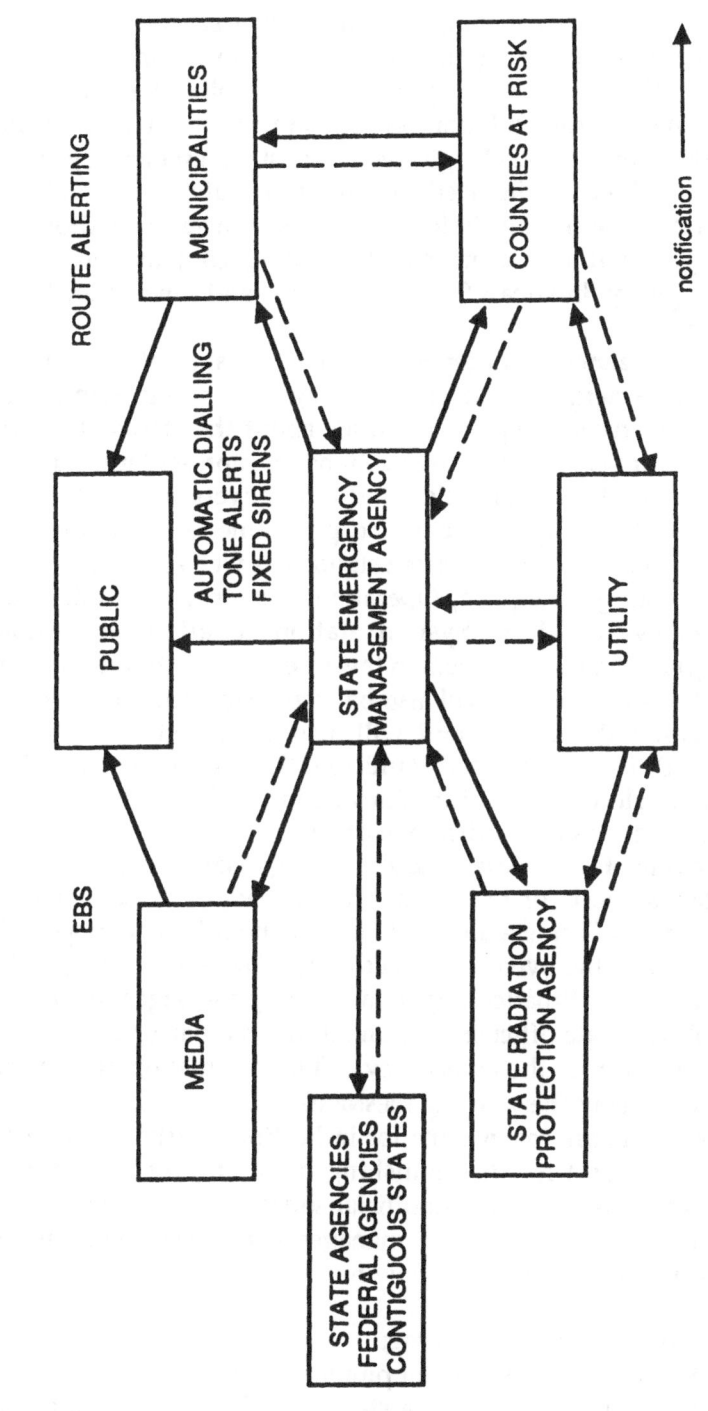

* Notification is by dedicated phone lines, unless indicated otherwise.

State emergency-management officials assume the primary responsibility for coordinating and monitoring interorganizational communications. They are also responsible for public alert and notification. Following recommendations from state radiation experts, the utility, and state and federal agencies, the state emergency-management agency will assess the need for initiating public alert and notification via sirens, automatic dialling systems, tone-alert radios, and route alerting. It will also coordinate the timing, content, and dissemination of Emergency Broadcast System messages and Emergency News Center releases.

An **emergency news center** will be established at the headquarters of the state emergency-management agency. Representatives of the media will receive regular briefings about the status of the plant, off-site conditions, and protective actions by representatives from the utility, state and federal agencies, and local government. This information will be sent directly, via dedicated phonelines and microwave systems, to Emergency Broadcast System stations as well as to other news media.

Each county Emergency Operations Center will be linked to the local Emergency Broadcast System station to allow activation of the Emergency Broadcast System in the absence of such directions from state officials. Risk counties will also be responsible for notifying municipalities within the county and will also be linked with each other by dedicated telephone lines. Municipalities will be responsible for siren and route alerting. Municipalities within a given risk county will have communications links with each other.

The role of the media. The state emergency-management agency and member stations of the Emergency Broadcast System will establish memoranda of understanding to clarify the roles and responsibilities of the media in the event of an accident. The state agency will provide the Emergency Broadcast System warning messages for direct relay to the public. State officials will undertake regular briefings by a variety of experts during the emergency. This information and interpretation will supplement the warning messages.

The media are responsible for indicating clearly the sources of their information and for distinguishing between official information and information derived from unofficial sources. Television stations will be provided by state emergency-management officials with on-line graphics displays of information showing:

- the location of the plume;
- the projected direction of plume movement;
- the location of screening stations and decontamination and mass-care centers; and,

- the traffic and weather conditions.

These graphics displays will be relayed to the public by the media in the event of an emergency. State officials and the media will practice this capability during drills and exercises.

Resiliency and redundancy. During a nuclear plant emergency, there must be continuous tracking of off-site conditions such as:

- the local weather and traffic conditions,
- problems at screening stations and mass-care centers, and
- the general state of public response.

A dense network of communications, both vertical and horizontal, among off-site emergency workers at all levels is needed to assure that accurate and timely information on off-site conditions is always available. The risk counties will monitor these conditions closely and issue regular status reports to the state emergency-management agency and to each other. This information will be relayed to emergency workers and members of the public as necessary to modify ongoing activities and adjust to new situations.

Multiple channels of communication and backup systems will be provided, including sufficient staff for additional shifts and replacement personnel. Dedicated phonelines connect the state emergency-management agency with the utility, other state agencies, and the risk and support counties. A computer network and backup radio communications supplemented these phonelines. Within 25 miles of the plant, dedicated phonelines will link the risk counties to each other, the municipalities, and the Emergency Broadcast Systems stations. Radio communications again will serve as backups.

17. How will the public be alerted and informed about the changing situation?

In the event of an accident at a nuclear plant, the public will receive information and advice through the radio and television stations of the Emergency Broadcast System (EBS). Members of the public will be alerted to tune to an Emergency Broadcast System station in different ways, according to the emergency planning zone in which they reside:

- The Inner Planning Zone will be alerted by the activation of tone-alert radios, fixed sirens, and route alerting.
- The Middle Planning Zone will be alerted by fixed sirens and route alerting.
- The Outer Planning Zone will receive information through the general media.

The warning system. The warning system at the nuclear plant has two functions:

- First, it will alert people that an emergency exists and they should tune into local Emergency Broadcast System television and radio stations.
- Second, it will provide additional information and guidance, with constant updates as the emergency unfolds.

The Emergency Broadcast System will serve as the primary source of information during an emergency, and this will be supplemented by additional information from the Emergency News Center. In the *Inner Planning Zone*, the utility will provide and maintain tone-alert radios for all residents within five miles of the plant. These radios will be tuned to a special local EBS station, and will be activated automatically in an accident. The Emergency Broadcast System will carry information bulletins about the status of the plant and recommended protective actions. Fixed sirens and route alerting will supplement the system of tone-alert radios. Those persons who fail to hear Emergency Broadcast System messages over the tone-alert radios (because they are beyond their reach or because the radio is inoperative) will be alerted to tune to a local Emergency Broadcast System station by sirens. The sirens will also serve as a backup system should the tone-alert radios fail, and vice versa. Route alerting will be used in areas of low population densities and poor siren coverage. The utility will also provide tone alert radios and automatic dialling systems to all firms with more than 50 employees, and institutions, such as schools, nursing homes, day care centers, hotels, and tourist facilities within five miles. This extensive alerting system, which may involve multiple means of communication (see Box), is justified by the higher risk and shorter response time that characterizes this area.

In the *Middle Planning Zone*, large employers and key institutions (e.g., schools, hospitals, prisons, day-care centers, nursing homes) and hotels, will be notified by automatic dialling systems, with tone alert radios as a backup. In areas of dense populations, sirens will be used to

---

EMERGENCY COMMUNICATIONS DEVICES

Tone-Alert Radios are radios that are preset to a special local Emergency Broadcast System station and will be activated remotely by state officials.

Automatic Dialling Systems are capable of calling multiple telephones at the same time with the same message. They will alert people to tune to their local Emergency Broadcast System stations.

Fixed Sirens will be placed at appropriate intervals within the inner and middle planning zones. A continuous, steady three minute tone will alert people to tune in to their local Emergency Broadcast System station.

Route Alerting is the use of mobile sirens, and loudspeakers fixed to cars and trucks, to alert people to tune to their local Emergency Broadcast System station.

---

alert people to tune to an Emergency Broadcast System station. Route alerting will supplement the siren system, especially where population is sparse and siren coverage poor.

The *Outer Planning Zone* requires no special alert and notification system, and residents of this area will not be required to take immediate action. Following most accidents the plume will take several hours to travel such distances, and the radiation levels will be considerably reduced due to deposition and dilution. Emergency responses may be necessary, however, especially in areas of heavy rainfall. The implementation of such responses will be based on environmental monitoring. The public will be informed via normal media channels.

Keeping the public informed. Members of the public will receive two kinds of information through the media:

- First, under agreement with state officials, the media will release official emergency messages. These messages will provide details about accident conditions and recommended protective actions (see Box below).
- Second, the media will receive regular briefings from the appropriate state agency (PEMA) about accident conditions. They will also seek from other expert sources (identified in previously established lists) further background information and perspective about the emergency as it unfolds. The emergency messages will be clearly distinguished as to source.

SAMPLE EMERGENCY MESSAGE

We interrupt our program to bring you the following message.

THIS IS NOT A TEST

An **ALERT** condition was declared at _____ (time) today at the _____ Nuclear Generating Station. There has been no release of radiation, but the Governor and emergency organizations have been notified in case conditions deteriorate.

At this time, if you live within 25 miles of the plant, you should:

- locate and read your emergency information brochure or calendar;
- turn to the emergency information section in your local telephone directory if you cannot find the brochure, or if you are away from home;
- remind yourself of your family emergency-response plan, and;
- stay tuned to this station for further information.

The Governor and emergency management officials are monitoring closely the conditions at the plant. If these conditions improve, we will inform you immediately. If these conditions deteriorate, the Governor will recommend what to do, and we will inform you immediately. No special precautions are necessary for school children at this time, and there is no need for you to collect your children. Please stay tuned to this station for further information.

Once again, the _____ Nuclear Generating Station declared an **ALERT** at _____ (time) today. There has been no release of radiation, but the Governor and emergency organizations have been notified in case conditions deteriorate. This message will be repeated in ten (10) minutes unless new information becomes available. Please stay tuned to this station for the latest official information.

The emergency information will indicate:

- the nature and cause of the accident, if known,
- the likely time available before a release,
- the location and direction of any radioactive plume,
- off-site traffic and weather conditions, and
- the recommended protective actions.

If an evacuation is required, state officials will provide guidance as to:

- the area to be evacuated,
- the most suitable routes to follow (see also Section 19),
- the location of screening stations (see Section 20),
- and the location of decontamination and mass-care centers (see Section 20).

This information will be portrayed graphically on television, using maps of the region.

Talking with neighbors. In any emergency people get information from sources other than the media and local officials. People talk to others at home, in the neighborhood, and at work. Emergency planners, decision makers, and emergency workers should recognize this and encourage people to talk to each other. Their goal is to provide clear, comprehensive information in a timely fashion, and thereby to minimize the generation and circulation of false and erroneous information. Rumors abound during emergencies. They are virtually uncontrollable after their generation, but their extent and impact can be minimized by providing a consistent and comprehensive flow of information from the beginning and throughout the duration of the emergency.

SHORT-TERM PROTECTIVE ACTIONS

18. What short-term protective actions can the public take?

Short-term protective actions* include:

- evacuation, which can reduce, to very low levels in favorable circumstances, direct exposure of all kinds;
- sheltering, which can reduce direct exposure of all kinds;
- respiratory protection, which can reduce direct exposure through inhalation;
- ingestion of potassium iodide, which can reduce thyroid exposure from inhalation of radioiodides; and,
- rapid interdiction of food and water supplies, which can reduce indirect exposure.

The information that follows is intended to provide emergency workers and the public with guidance about actions that may reduce potential harm. This guidance will also be included in pre-emergency educational materials as well as in emergency broadcasts during an accident.

Evacuation. The most effective means of reducing direct exposure to radiation is for people to evacuate before the radioactive plume reaches them. If such precautionary evacuation is possible, exposure to the initial plume can be eliminated altogether (subsequent low levels of exposure cannot be prevented because the accident will produce widespread contamination).

In practice, evacuation will often not occur so promptly. The accident may develop so quickly that warning time is inadequate for precautionary evacuation. Also, heavy traffic or adverse natural conditions

---

*In this plan, short-term protective actions are defined as those actions that are initiated within 48 hours of a release (or the major part thereof). This time period is shown in Figure 12.

FIGURE 12

Timing of Short-Term and Long-Term Protective Actions

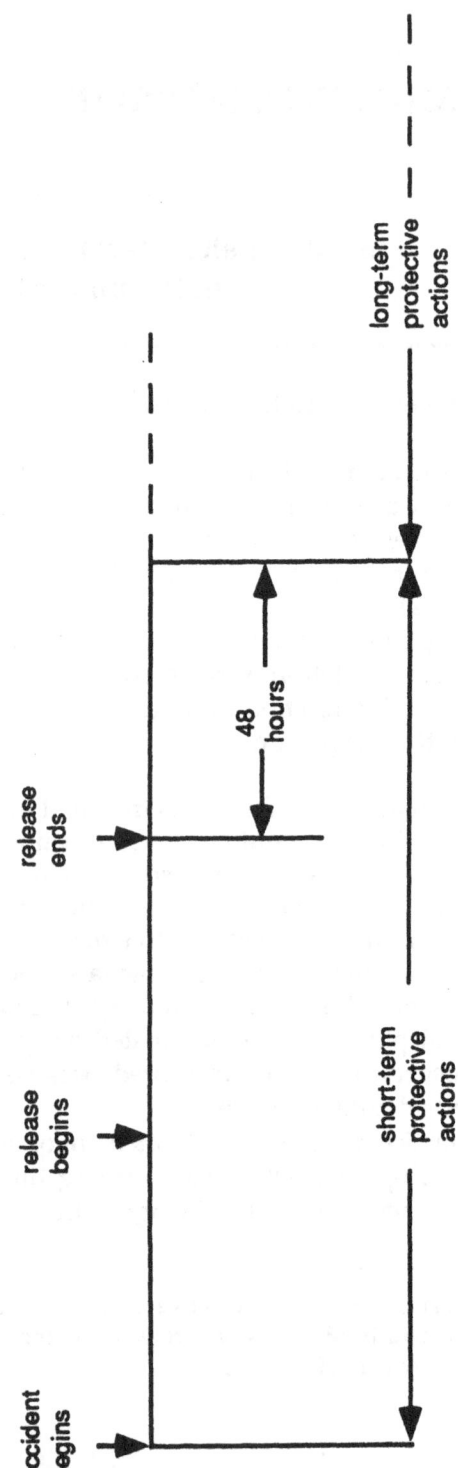

(e.g., heavy snowfall) may hinder an evacuation. Thus, sheltering and other protective measures may play an important role.

If asked to evacuate people should:

- gather together family members, if nearby;
- stay tuned to the local Emergency Broadcast System station for further instructions and traffic conditions;
- if directed to do so, go to the nearest screening station, even if it seems that exposure is unlikely;
- after screening, and decontamination if necessary, go to a friend, relative, or mass-care center outside the Middle Planning Zone;
- assist neighbors without cars; and,
- call the local Emergency Management Agency if no means of transport is available.

Sheltering. Sheltering reduces exposure by placing solid material between people and the radiation source. At best, a person who finds good sheltering conditions—in a multi-story concrete building, for example—might be exposed to 1/5 the intensity of radiation from the plume and 1/10 the intensity of exposure from radioactive material deposited on the ground as compared with a person standing in the open. Likewise, a well-sheltered person would receive an inhalation dose about 1/5 that of a person in the open (see Table 7). If advised to shelter, people should:

TABLE 7

Illustrative Dose-Reduction Factors* for Various Shelters

Type of Shelter	From a Passing Plume	From Material Deposited on the Ground	From Inhalation Exposure
Small frame building without basement	1	2	2
with basement	3	5 to 10	3
Multiple-story concrete structure	5	10	5

* Unshielded dose divided by the reduced dose.

- go inside and stay there;
- move to the lower levels of a concrete building;
- go into a deep basement if one is available;
- if a basement is not available, sit or lie on the lowest level of the building, as far away from the walls as possible;
- close tightly all doors and windows, to reduce the infiltration of radioactive materials;
- turn off fans and air conditioners;
- keep pets inside and shelter farm animals (if possible);
- stay indoors until they receive official notice that it is safe to go out, and
- if they are in a car, close the windows and air vents and shut off fans and air conditioners.

Respiratory protection. If in the open or in a relatively leaky building, people may protect themselves by breathing through a crumpled handkerchief or bath towel. A crumpled handkerchief can reduce inhalation exposure to 1/10 (10%) of the level without protection (Table 8).

In the absence of good shelter and/or if advised to use respiratory protection, people should:

- seek adequate shelter as soon as possible, and follow directions for sheltering;
- hold a crumpled handkerchief, towel, or other piece of material over mouth and nose;

TABLE 8

Respiratory Protection (Against Particles) Provided by Common Household and Personal Items

	Number of Thicknesses	Efficiency of Reduction (%)
Toilet paper	3	90
Handkerchief, man's cotton	Crumpled	90
Bath towel, Turkish	1	70
Bed sheet, muslin	1	70
Handkerchief, man's cotton	1	30

- tie handkerchief or other material in place, or use surgical masks if available;
- aid children and others in need of assistance in following these instructions; and,
- keep protection in place until official notice that plume has passed.

Respiratory protection of this kind has its primary value during passage of the plume. Since people will be unable, without instruments, to detect the presence of the plume, it is wise to continue using respiratory protection until official notice is received that the plume has passed.

Potassium iodide for thyroid protection. The thyroid gland is unique in that it can be protected from radiation exposure through medication. Part of the radioactive release will comprise radioactive iodine, which, if inhaled, will pass through the bloodstream to the thyroid gland. The radioactive iodine will concentrate there and could cause thyroid damage. This can be prevented if nonradioactive iodine is ingested prior to or soon after the arrival of the plume.

Nonradioactive iodine to protect the thyroid gland is packaged in the form of tablets or a solution of potassium iodide (KI). If advised to use potassium iodide, people should:

- locate protective action kits containing supply of potassium iodide;
- follow carefully official instructions about who should use KI, when to take it, and in what doses; and,
- aid children and others in need of assistance.

Within 25 miles of the plant, kits containing potassium iodide and a surgical mask for respiratory protection will be predistributed to all persons. These kits will also be stockpiled (at the expense of the utility) at schools, fire and police stations, and other public facilities. Beyond the 25-mile zone, utilities will inform their customers that these kits will be available for a nominal charge, from the State Department of Public Health. Finally, citizens nationwide may at any time purchase supplies of potassium iodide from pharmacies and state health departments (see Box).

Reducing skin contamination. Skin contamination can be reduced by minimizing the exposure of the skin and by changing clothes carefully after the plume has passed. If advised to decontaminate themselves, people should:

POTASSIUM IODIDE (KI) FOR THYROID PROTECTION

Potassium iodide (KI), taken by tablet or as saturated solution, can protect the human thyroid gland from radioactive iodine released into the air. Public health officials will issue recommendations as to which people should take KI and when.

A person advised to take this medicine should take one dose immediately and another dose each 24 hours thereafter. It should not be taken more often, however, because this will not help and may increase the risk of side effects. KI should be taken for a total of 10 doses or until public health authorities issue advice to cease taking this medicine.

Within 25 miles of the TMI nuclear plant, the state emergency-management agency (PEMA) will be responsible for stockpiling emergency kits containing KI at schools, fire and police stations, and other public facilities. Electric utilities will pre-distribute such kits to all households in the Inner and Middle Planning Zones, and will inform customers else-where in the state that the kits may be obtained from the Department of Public Health. Finally, citizens nationwide may at any time purchase supplies of KI from pharmacies and state health departments.

In a nuclear accident, KI should be taken only in the form packaged for use during a radiation emergency. Detailed instructions will be provided with this package and should be followed closely. The normal doses are:

Tablets: (130 mg of KI per tablet)
ADULTS AND CHILDREN 1 YEAR OF AGE OR OLDER:
One (1) tablet once a day. Crush for small children.
BABIES UNDER 1 YEAR OF AGE:
One-half (1/2) tablet once a day. Crush first.

Saturated Solution: (21 mg of KI per drop)
ADULTS AND CHILDREN 1 YEAR OF AGE OR OLDER:
Add 6 drops to one-half glass of liquid and drink each day.
BABIES UNDER 1 YEAR OF AGE:
Add 3 drops to a small amount of liquid once a day.

WARNING

Potassium iodide should not be used by people allergic to iodine. Keep out of reach of children. In case of overdose or allergic reaction, contact a physician or the public health authority.

- remove clothes carefully and place in plastic bags;
- shower thoroughly, paying particular attention to hair and body orifices;
- avoid scratching skin;
- dress in clean clothes, including shoes; and,
- place contaminated clothing outside living quarters.

Rapid interdiction of food and water supplies. Public health authorities will act promptly to prevent the movement of milk, water, or agricultural produce from contaminated areas. These actions will constitute the first steps in a series of long-term protective actions, as described in Section 24 below.

The public will be warned if the area has been so contaminated that surface water and home-produced milk or produce cannot be consumed. Several days may pass, however, before accurate monitoring information is available for each potentially affected area, so the public needs to be alert to this potential problem.

Recommendations by public authorities. Public authorities will issue recommendations stating protective actions that are needed for people living at various distances and directions from the plant. Especially people living within the Inner Planning Zone (five mile-radius) need to follow such recommendations immediately. People at all distances, however, should treat the recommendations seriously. Pregnant women, parents, and teachers should be particularly careful to take all recommended protective actions, because fetuses and children are especially susceptible to radiation exposure.

---

19. If an evacuation is required, how will it be organized?

- The Governor will order the evacuation, on the advice and recommendation of state emergency-management officials (PEMA).
- These officials will relay this decision to counties, municipalities, and the public through the emergency warning system (see Sections 16 and 17).
- They will also provide, via television and radio, continuous guidance on evacuation routes, plume movement, screening stations, and traffic conditions.

- The officials will mobilize screening stations and decontamination and mass-care facilities on all major routes leaving the evacuation area.
- Members of the public will choose their own destinations and evacuation routes in light of the guidance provided by the state officials.
- Institutions with special populations will implement approved evacuation plans.
- The state officials will continuously monitor the evacuation, through local officials and airborne surveillance, making adjustments where needed.

Evacuation. As noted previously, the primary objective of emergency planning and response is to minimize harm to life and to keep any doses below Protective Action Guides and otherwise as low as is reasonably achievable. The most effective means of reducing direct radiation exposures in the event of a radiation release is for people to evacuate before the plume reaches them. If such preemptive evacuation is possible, initial exposure to the plume can be eliminated altogether. In some unusual cases, the accident may develop so rapidly that evacuation is impossible, or other conditions, such as a major snowstorm, may impede an evacuation. On the other hand, experience suggests that evacuations can be successfully conducted, even in the face of adverse conditions.

Roles and responsibilities. Only the Governor has the authority to order an evacuation. The decision to evacuate will be made by the Governor, on the recommendation of state emergency-management officials. These officials will consult with the utility and the state radiation agency about the likely accident characteristics, such as the time available before a release, the composition, size, duration, and direction of the plume; and the possible doses at different distances. The agency will use this technical evaluation to determine whether an evacuation is necessary, and feasible in the time available.

To aid in this decision, evacuation time estimates are available that indicate the expected time necessary to evacuate the population around the plant under different conditions. Evacuation could well take longer if initiated during rush hour or during a major snowstorm. State officials will determine how large the evacuation zone should be. The most intensive emergency-planning and -preparedness effort focuses on the Inner Planning Zone, because rapid evacuation is particularly important for this zone. Many accident scenarios, however, will require evacuation beyond five miles.

The decision to evacuate is relayed to emergency-response organizations and members of the public through the warning system (see

Sections 16 and 17), which will alert members of the public to tune to a local Emergency Broadcast System television or radio station for further details. Emergency messages will explain why an evacuation has been recommended, how it will be organized, and how evacuation will avoid the moving plume.

Evacuation routes. To maintain flexibility in evacuation, emergency planners will predetermine evacuation routes but will not predesignate specific routes for the public. Police and emergency workers will be assigned to these predesignated routes. As information becomes available about weather and traffic conditions and the likely direction of plume passage, it will be relayed to the public via television and radio. On the basis of this information, public officials will provide guidance as to the best evacuation routes. But members of the public will be free to choose their own evacuation routes in light of this information and their preferred destinations. State officials will mobilize emergency personnel to direct traffic, assist with breakdowns, provide emergency fuel, and maintain priority for emergency vehicles according to arrangements stated in approved county and municipal plans.

Screening and decontamination. In the event of evacuation, state emergency-management officials will activate screening stations on all major routes leaving the evacuation area. The location of these stations will vary according to the nature of the accident and the prevailing weather conditions. Local television and radio stations, augmented by the use of traffic guides, will direct people to be screened for exposure and will provide the locations of screening stations. Those people with measurable radioactive contamination at the screening stations will be directed to the nearest decontamination center (for more details, see Section 20).

Special populations. Arrangements have been made for the evacuation of special populations, including:

- school children
- college students
- hospital patients
- nursing-home residents
- prison inmates
- the handicapped
- individuals lacking private transport

These arrangements include:

- preparing institutional emergency plans;
- maintaining lists of handicapped individuals and their needs;

- providing special alert and notification systems (such as direct-dialling and tone-alert systems); and,
- developing special transport (including the provision of ambulances and buses).

Hospitals, nursing homes, and prisons will develop institutional plans and make appropriate contractual arrangements for transportation. These institutions will also establish, with similar institutions outside the Middle Planning Zone, written agreements to serve as hosts in the event of an evacuation. Schools and day-care centers will also prepare institutional emergency plans and transport arrangements. If sufficient time precedes an evacuation, children will be sent home as they would during a snow emergency. It is expected that some parents will collect their children in an emergency, and the remaining children will be bused to predesignated mass-care centers. Parents will be reminded biennially which mass-care center has been designated. Teachers will remain in control of their classes until parents pick up their children either from school directly or from the mass-care center. Teachers will be responsible for maintaining and updating accurate student rosters.

Colleges will develop emergency plans outlining roles and responsibilities. Students with cars will be allowed to leave for their own destinations. Those without cars will be bused to predesignated mass-care centers. Municipalities will keep an updated inventory of the handicapped requiring special transport and of those without private transport. Municipal and county governments will also provide suitable transportation to evacuate those individuals.

---

20. What will be the procedures and resources for screening and decontamination?

In the event of an accident requiring an evacuation:

- screening stations will be established along major routes of evacuation to identify those individuals with radioactive contamination and/or other indications of significant exposure;
- decontamination centers will be established outside the evacuated area to decontaminate all those individuals with significant levels of contamination;

- mass-care centers will be co-located with decontamination centers to provide accommodation for individuals after decontamination and for others with no alternative lodgings.

For all evacuations, including those of a precautionary nature, the state emergency-management agency (PEMA) will mobilize screening stations and decontamination and mass-care centers at appropriate sites around the designated evacuation area. In the event of a precautionary evacuation, prior to any radiation release, screening and decontamination will not be necessary, but mass-care centers will be open to those without alternative lodgings. In these cases, personnel at the screening stations will aid those evacuating by directing traffic and giving directions to nearby mass-care centers.

This early mobilization of resources is necessary to allow for the rapid implementation of screening and decontamination should a release occur during a precautionary evacuation. In the event of an evacuation following a release, screening stations, and decontamination and mass-care centers will be mobilized and activated as quickly as possible. The remainder of this section of the plan deals specifically with the evacuations involving a release of radiation.

One cannot overestimate the importance of screening and decontamination. Since radiation is colorless, odorless, and tasteless, only screening will identify those individuals who have been exposed. Screening will also indicate the degree of exposure and the necessity for decontamination and medical care. Local television and radio stations will advise members of the public of the necessity for screening and the location of screening stations. The state emergency-management agency will establish a sufficient number of screening stations at appropriate sites along all major routes leaving the designated evacuation area. The use of traffic controls will minimize traffic delays. The stations will be located sufficiently far away to avoid causing traffic to back up into the designated evacuation area. Numerous sites within the region will be predesignated for the location of screening stations and decontamination and mass-care centers. Selected sites will be activated according to the nature of the accident and the configuration of the evacuation area. This system of predesignated sites will allow greater flexibility in dealing with the variety of accident scenarios and changing accident conditions. The goal is to screen all evacuees, but some evacuees may well arrive at their destinations without having gone through the screening and decontamination process. Emergency Broadcast System broadcasts will provide these people with guidance as to how they may obtain screening for possible contamination and for emergency procedures that assist in personal decontamination (e.g., discarding clothing, showers, etc.).

At the screening station. While waiting in line at the screening station, individuals will be asked about any health symptoms, their point of departure, and their intended destination. Screening teams will identify people who may have been exposed but who show no signs of contamination.

People who may have been contaminated or exposed will be given an identification tag and directed to the nearest decontamination center. Others will continue on to their original destinations. Persons who have no friend or relative to go to will be directed to a mass-care center. Figure 13 shows the overall structure of procedures that will be used for decontamination.

At the decontamination center. Schools and colleges beyond the Inner and Middle Planning Zones will be predesignated as decontamination and mass-care centers. Only some of these will be activated in an emergency, depending upon the size and shape of the evacuated area. Since several screening stations will direct people to a single decontamination center, there will be fewer decontamination and mass-care centers than screening stations.

Each decontamination and mass-care center will screen people arriving without tags from screening stations. These people will receive tags according to their degree of contamination or likely exposure as well as advice on how to proceed. Those arriving with tags from other screening stations will receive similar guidance.

Contaminated cars will be parked in a large open space adjacent to the center and washed down with fire hoses. Screening and washing will be repeated until the cars are free of significant levels of radiation. People whose cars have been sufficiently decontaminated will continue on to their original destinations. People whose cars cannot be sufficiently decontaminated will not be allowed to leave in them (these cars will be impounded), and they will remain at the mass-care center.

An individual who has been contaminated or who may have been exposed will walk to a large tent that will serve as the mobile decontamination station. There will be separate tents for males and females. Here the person will undress and put all clothing, which will not be returned in plastic bags. People will be instructed in washing techniques and directed through gang showers. After the shower, each person will be toweled dry and re-screened by Geiger counter. While screening, decontamination personnel will again review information on symptoms and travel routes.

FIGURE 13

Screening, Decontamination, and Mass Care

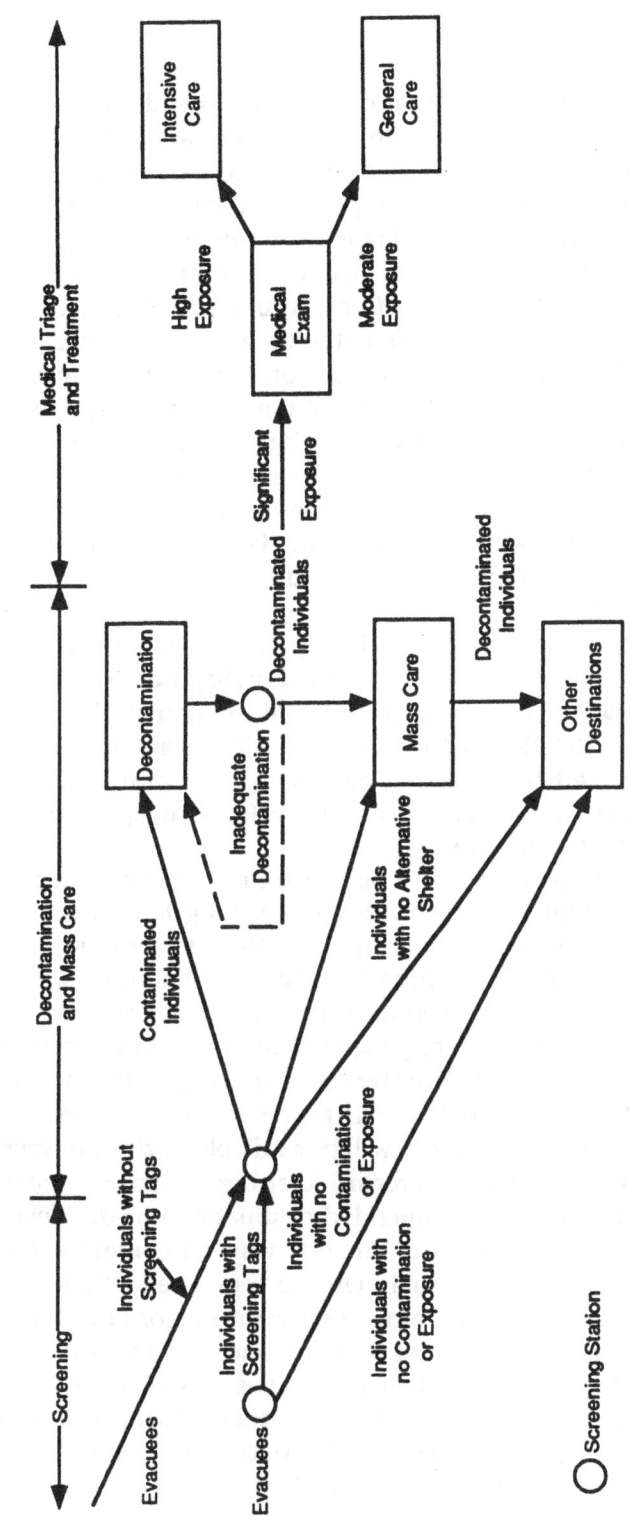

People with no residual contamination and with no other indication of exposure will proceed to the dressing area. Here they will receive disposable clothing as well as a color-coded tag (indicating the level of decontamination). Then they will be escorted out of the decontamination tents into the adjacent mass-care center.

People who have some residual contamination will be redirected through the showers. If, after a second shower, there are still areas of radioactivity on a person's body, an assistant trained in decontamination techniques will help the person remove the remaining radioactivity. Once decontaminated (and if there are no other signs of exposure), these people will receive a color-coded tag, dress in disposable clothing, and proceed to the mass-care center.

At the mass-care center. At the mass-care center, clerks will register each person, noting the tag color. Food, blankets, chairs, beds, bathrooms, television sets, and radios and other amenities will be available.

Contaminated or exposed people. People with signs of moderate or substantial exposure will be dressed in disposable clothing and assigned to an intake booth adjacent to the decontamination area. On the basis of a more detailed history and a physical exam, medical personnel will decide whether to send a person to a general hospital (for patients with moderate exposure) or to a referral hospital (for patients with more serious exposures).

Medical personnel will be in continuous communication (either by radio or cellular phone) with the receiving hospitals, in order to coordinate patient flow. Vans will take the less seriously exposed to their assigned hospitals. Where necessary, ambulances will take the more seriously exposed to their destinations. As the regional and state referral centers for the severely exposed fill up, it may be necessary to transport patients by helicopter or fixed-wing aircraft to out-of-state hospitals. State officials will be responsible for ensuring that these resources, and procedures, will be available during an emergency.

Communications and record-keeping. Prior to transfer, patients will be registered and their intended destination noted. Next of kin will be notified of their location. In this way all contaminated and significantly exposed individuals will be registered. State officials will be responsible for maintaining registration records in an accessible and retrievable form, including standard and computerized formats.

The state emergency-management agency will also be responsible for ensuring effective communications among hospitals and medical triage sites. The equipment needed (radio, cellular phones, and so forth) will be identified and planned for in advance.

Organization and training. Emergency medical personnel, including volunteers, will be trained to maintain a flow of information to people moving through the decontamination site to the mass-care center. State emergency-management officials will assure appropriate training and exercising of the various roles and functions. Personnel involved in screening, decontamination, and medical treatment will need specialized training in radiation monitoring and the treatment of radiation health effects (see Section 28).

Resource requirements. *Screening stations* will involve a road block and sufficient trained personnel to take personal histories and conduct rapid Geiger screens.

Each *decontamination and mass-care center* will have a screening area, a decontamination area that includes showers, and an adjacent mass-care center. A large open area to serve as the car parking lot will be needed.

Mass-care centers will be located in existing municipal buildings, such as schools, which have facilities to feed and shelter large numbers of people. Large tents adjacent to these buildings will be used for screening and decontamination. Facilities for water supplies to the showers will be needed. (This plan does not call for a system for collecting run-off, since this will involve a very large volume of minimally contaminated water).

The state emergency-management agency is also responsible for stockpiling supplies of disposable (paper) clothing, plastic bags for storage of contaminated clothing and other items, and towels, soap, and other materials needed for decontaminating large numbers of people.

Vans for patient transport will be identified in advance from the civilian supply as potential call-up vehicles. They and their drivers will be expected to arrive, as planned, at the correct site when and if the emergency is announced. Ambulances assigned to each site will provide patient transport as needed. Air-transport facilities and vehicles will also be arranged in advance. During an emergency they will be placed on standby notice.

The actual stocking, equipping, and staffing of the mass reception centers will be carried out by agencies of the local emergency medical system in cooperation with other municipal, county, and state agencies and trained volunteer personnel.

21. How will medical treatment be organized?

In the event of a nuclear accident, the emergency medical system will involve three tiers of preplanned mobilization:

- The first tier will be the physicians, nurses, laboratory technicians, and other health-care workers who will staff the decontamination and mass-care centers;
- the second tier will be the hospitals, located outside the Middle Planning Zone, that will receive people requiring medical treatment and follow-up care; and,
- the third tier will be state, regional, and national hospitals with intensive-care units capable of dealing with severe radiation injuries and illnesses.

The hierarchy of medical care will require a significant effort and sufficient resources to assure preparedness. Participating hospitals will develop appropriate institutional plans and contractual arrangements. State officials will be responsible for assuring the effectiveness of this system, and its functions will be activated upon the directive of the Governor.

Functions. The emergency medical system will have four major functions:

- to identify those contaminated with radiation and/or those with other evidence of exposure;
- to provide decontamination and medical care as needed;
- to monitor exposed and contaminated individuals after initial identification; and,
- to maintain accurate medical records of all contaminated and exposed individuals.

Hierarchy of medical care. To achieve these functions, the medical system will be organized into three tiers of care. The first tier comprises the physicians, nurses, laboratory technicians, and other health care workers, at the decontamination and mass-care centers. These personnel will be responsible for identifying contaminated or exposed persons, and referring them for further appropriate medical care. Screening stations, separate from decontamination and mass-care centers, will be operated by trained personnel, but not medical staff, who will be in limited supply.

People needing medical attention will be referred to designated hospitals for treatment and follow-up care. These hospitals form the second tier of the medical system. They will be located outside the evacuation zone but close enough to facilitate the transport of patients by cars and ambulances. An important subgroup of such patients will be those emergency workers and perhaps members of the public who have received nonradiological injuries requiring prompt treatment and who are also contaminated with radioactivity.

The third tier comprises state, regional, and national hospitals with intensive-care units, including burns units, capable of dealing with severe radiation injuries and illnesses. These hospitals will receive those individuals with the most serious radiation exposures. Hospitals capable of handling such patients are relatively few. Hence, a national response effort, including transporting such patients by air, may be necessary.

Organization and resources. To be successful, this hierarchy of medical care will require advanced planning and preparedness. The accident at Chernobyl illustrated the extraordinary level of medical resources that may be required in the event of an accident (see Box). The state emergency-management agency will assure that participating hospitals draw up institutional plans and make appropriate contractual arrangements for communications, transportation, and the provision of services. Medical staffs will receive specialized training

---

MEDICAL MOBILIZATION AND RESOURCE NEEDS DURING THE CHERNOBYL ACCIDENT

The Soviet medical response to the accident at Chernobyl entailed an extraordinary mobilization of resources at a national level. The response was prompt: within five hours more than 100 people were hospitalized; within 24 hours teams of specialists had arrived and examined more than 350 people and performed more than 1,000 blood analyses. In all, 1240 physicians, 920 nurses, and well over 3000 other assistants were mobilized. More than 200 people received severe radiation exposures; they were treated in specialized hospitals in Moscow and Kiev. After the accident, some 18000 people reported to hospitals for checkups. Since most of the radioactive material was lifted high in the air, the primary people receiving life-threatening exposures were plant personnel, fire fighters, and other emergency workers.

in their roles and responsibilities. Hospitals must be prepared to decontaminate patients and prevent the spread of radioactive contamination.

The mass-care centers will be mobilized by the redeployment of
local physicians, nursing staff, and other health care workers. These
people will require special advance training in screening,
decontamination, health effects, and record keeping. The hospitals in
the second tier will establish plans to discharge less needy patients
and reassign staff in the event of an accident. Personnel will receive
training in health effects and medical care and the institutional
arrangements for communications and referral.

The third tier will comprise the major teaching hospitals with
intensive-care units. Personnel will need training in the institutional
arrangements for receiving accident victims. They may need additional
training in the nature of power plant accident exposures, health effects,
and suitable medical care.

Arrangements for communications and the transportation of patients
will be carefully established in advance of an accident. Lines of authority and communications among decontamination and mass-care
centers and referral hospitals must be clear and should be exercised
periodically.

---

22. What about school children, college students, hospital patients, and other special populations?

Populations in the nuclear plant region that will require special
attention include school children, college students, hospital patients,
nursing home residents, prison inmates, handicapped persons, ethnic
and cultural groups, people without private transport, and transients
(people temporarily in the area on business or vacation). To meet their
needs during a nuclear emergency:

- Special provisions will made for these populations, including
  enhanced warning systems, targeted information, and additional
  transportation arrangements;
- each institution will draw up and regularly exercise an institutional emergency plan;

- the emergency plans for day-care centers and public and private schools will include special provisions for warning, evacuating, and sheltering (see Sections 17, 18, and 19); and,
- municipalities will be responsible for maintaining accurate lists of handicapped persons and those lacking private transport and for providing protection during an emergency.

Goals. The goal of special provisions for these populations is to assure that their particular needs will be identified in any emergency response and that they will be afforded protection commensurate with that of other populations.

School children and college students. School children and college students pose a number of special problems. These include:

- greater susceptibility to radiation-induced health effects;
- less capability to take protective action; and
- separation from other family members.

To assure timely notification and the implementation of protective actions, the following provisions and procedures will be instituted:

- The decision to adopt any special protective action for school children and college students will be made by the Governor upon a recommendation from state emergency-management officials and in consultation with the state department of education.
- All schools and colleges in the Inner and Middle Planning Zones will be linked to state officials and county Emergency Operations Centers by automatic telephone-dialling systems. School officials will be alerted by tone-alert radios and by telephone calls from both the state emergency-management agency and county officials. Fixed sirens and route alerting will serve as backup alerting systems.
- Each school, school district, and college will develop its own emergency plan outlining responsibilities, procedures (including destination points and busing arrangements), and other protective actions. These plans will be developed in consultation with and reviewed by the appropriate counties and state officials.
- School officials should recognize that some parents will arrive to take their children out of school and will have plans to facilitate those actions without compromising their overall emergency-response capability.

- Protective-action kits, including potassium iodide, surgical masks, and instructions, will be stockpiled in all schools and colleges in the Inner and Middle Planning Zones.
- Municipalities in the Inner Planning Zone will design and implement special training programs for teachers and educational programs for school children.

Other institutional populations. Hospitals, nursing homes, day-care centers, homeless shelters, and prisons in the Inner and Middle Planning Zones will have the same communications and alerting system as schools and colleges. They also will develop their own emergency plans and contractual arrangements necessary for implementing protective actions. Personnel in these institutions will receive special training in emergency-response procedures. Proper use (and potential modifications) of ventilating systems while sheltering will be addressed. Prisons will require special plans for handling security issues and must arrange destination facilities in advance.

Handicapped persons. Handicapped persons, such as the hearing-impaired, will require special emergency-preparedness arrangements, such as teletype machines for alert and notification. Mobility-impaired individuals and those persons lacking private transportation will require special transportation arrangements in the event of an evacuation. Each municipality will maintain, and regularly update, a list of such persons, their locations, and needs. County officials will work with municipal personnel to develop means to meet these needs in an emergency. Special contractual arrangements for transportation will be developed as part of the county and municipal emergency plans.

Cultural groups. In Pennsylvania, for example, the Amish require special alert and notification procedures because their religious beliefs and customs prohibit the use of television sets and radios and because they live mostly in isolated rural communities. Many of the Amish did not learn about the TMI accident in 1979 until it was over. Because most of the Amish are engaged in farming and other rural pursuits, they will need special information and educational materials about the protection of animals, agricultural products, and water supplies (see Section 23). The most appropriate method of alert and notification will be route alerting using the Amish volunteer fire services. County and municipal governments will coordinate these activities through local Amish leaders. Informational and educational materials will also need to be channelled through local Amish leaders, and state and local governments need to develop closer relationships with the Mennonite Disaster Service, which has an extensive network of contacts in the Amish community.

Transients. Transients, such as tourists, migrant workers, homeless people, and people on business in the area, will be informed about emergency plans through special educational materials placed in hotels and public places. They will be alerted in the same way as the general population, and through special arrangements developed in hotels, soup kitchens, places of employment, etc. (see Section 17 for more details).

---

23. What are the special needs and responsibilities of farmers?

Farmers have special needs during a nuclear power plant accident. Within the Inner and Middle Planning Zones they will:

* be designated as emergency workers;
* receive special protective action kits; and
* receive special information and training in emergency preparedness and response.

Farmers also have special responsibilities:

* to protect themselves, their families, and their animals;
* to protect the public milk and food supply; and,
* to prepare a farm emergency plan in advance of an accident.

The information that follows is intended to provide farmers with guidance about actions that may reduce or mitigate potential harm. This guidance will also be included in pre-emergency educational materials as well as emergency broadcasts during an accident.

Contamination problems. In a nuclear accident, contamination of the milk supply is a major concern. Cows graze over large areas and can consume large amounts of radioactive materials deposited on the ground. Drinking contaminated milk may result in high individual radiation doses, especially in children. Contamination of meats, such as pork and beef, can also be a concern, although the time between slaughtering and consumption is longer. Contamination of other animal products, such as eggs and poultry, is of less concern because of the delay between exposure and consumption. Also, chickens are usually kept in closed quarters and fed on stored grains, and so receive less exposure.

Contamination of fruit and vegetables will be a problem only if an accident occurs immediately before or during harvest. In this case, fruit and vegetables should be stored until the radioactivity declines to tolerable levels, or the products should be destroyed. Farmers and members of the public will be warned not to consume fresh local produce.

Protective actions. If evacuation or sheltering is recommended, farmers should:

- immediately evacuate or shelter with their families.

If time permits, farmers should:

- remove dairy cows from pastures and put them on stored feed and a protected water supply in secure barns;
- milk cows before evacuating or taking shelter;
- avoid overcrowding animals and ensure adequate ventilation; and
- provide an adequate, protected water supply.

To meet these responsibilities, farmers will receive protective action kits, information, and training.

- **Protective-action kits** (including potassium iodide, surgical masks, flashlights, and dosimeters) will be predistributed to all farmers within the Inner Planning Zone and will be available at distribution centers in the Middle Planning Zone. Farmers will need the dosimeters to monitor their exposures if they stay in the evacuation area or if they have to reenter the area to tend to their animals. Farmers may also have to go outside to tend to their animals even though sheltering is recommended.
- **Special advance training** (see Section 27) will include the introductory program (giving background information on nuclear accidents, radiation, health effects, and protective actions) and a special program for farmers. These materials will provide additional details on preparing an emergency plan for the farm, including specific protective actions for the farmer and his or her family, animals, and crops. This training will be given annually to all farmers within the Inner and Middle Planning Zones.
- **Special information brochures** will be distributed annually to all farmers within the Inner and Middle Planning Zones. These will supplement the information brochures and calendars distributed to the general public. There will be no special provisions for alerting farmers aside from those that are used to alert the

general public (see Section 17). Local radio and television stations will, however, carry special information, advice, and official recommendations for farmers. The Governor, in consultation with state emergency-management officials and the state department of agriculture, will issue these recommendations and guidance.

LONG-TERM PROTECTIVE ACTIONS

24. How will food and water be protected?

For many people in the Outer Planning Zone, eating or drinking radioactive materials could be a source of significant radiation exposure following a serious accident. A number of actions may be needed to protect against such exposure:

- The US Department of Agriculture and the state department of agriculture will issue recommendations to the public on means to protect food supplies during and following an accident;
- the US Environmental Protection Agency and the state department of environment will issue recommendations to the public on means to protect water supplies during and following an accident;
- the state department of agriculture will establish and implement a monitoring program to test agricultural products for potential contamination and report findings to state radiation officials, other state agencies, county and local officials, and the public; and
- the state department of health, in consultation with state radiation and agriculture officials will determine, and implement, any directives for withholding or destroying contaminated agricultural products or for taking other necessary remedial actions.

Protection of food and water sources during the accident: State officials will recommend actions to protect food and water during an accident. These may include:

- covering cisterns and small gardens;
- bringing cows and other stock into barns, if feasible;
- using stored feed in place of grazing; and,
- harvesting or covering some crops before radioactive materials arrive.

Tracking the radioactive plume. The utility, state agencies, and federal agencies will monitor the passage of the plume and the distribution of deposited radioactive materials. Based on this information, state agriculture officials, with technical support from state radiation officials and federal agencies, will establish monitoring stations in and near potentially contaminated areas to test agricultural products for radioactive contamination. These stations will use state, local, and other approved independent testing facilities and personnel. As soon as personnel and equipment are available, state agriculture officials will organize teams to visit each potentially contaminated farm and to assess the degree of contamination.

Monitoring food and water. At each monitoring station, radiation measurements will be taken of drinking water and agricultural products. Whereas isotopes of cesium and iodine are expected to be the principal radioactive materials present, some samples will be examined to identify other possible radioactive elements. Levels of radioactivity will be compared to predetermined action levels, with the purpose of keeping exposures through ingestion well below the Protective Action Guides.

Milk contaminated with radioactive iodine may be processed (to cheese or dry milk) and used later after the radioactivity has largely disappeared. Milk, meat, and produce with excessive amounts of radioactive cesium, which remains radioactive for many years, will have to be destroyed. The monitoring stations will test the efficacy of washing various local vegetables and fruits before recommending washing as a sufficient protective measure. The monitoring stations will record their findings and systematically report to state radiation, health, and agriculture officials the results and the place of origin of contaminated material. Places of origin of uncontaminated produce will also be noted.

Contamination of land and crops. Teams measuring long-lived radioactivity in soils and in standing crops will report back so that recommendations may be made for:

- leaving things as they are;
- plowing to dilute mild surface contamination;
- scraping of surface layers of soil for more serious contamination;
- suspending use of the land if it is very seriously contaminated;
- cutting and disposing of contaminated standing crops.

25. How will relocation decisions be made?

A severe nuclear accident that results in substantial off-site contamination could require the relocation of some populations, based on comparing potential long-term exposures to long-term protective action guides (PAGs). If such action is required, the decision to relocate would:

- use *short-term protective action guides* to determine how quickly a population must move away from any contaminated area.

Any decision to permit the relocated population to return permanently to the area would:

- be based on the application of *long-term protective action guides.*

The decision to relocate. This decision will be taken by the Governor, on the advice of state emergency-management, health, and radiation officials, and will be based on long-term protective action guidelines. Long-term protective action guides are *annual* limits on exposure (expressed as doses per year) that are used to determine the suitability of land and buildings for residential and occupational use. Short-term Protective Action Guides are used to determine how quickly the population must move away from a contaminated area. (See Section 6 for a description of these guides). Relocation of exposed populations could occur in three phases, depending on the severity of exposure. When exposure rates are lower, more time can be allowed for organizing the relocation. This can facilitate better arrangements for temporary living quarters, removal of decontaminated property, and better assessment of the possibilities for decontamination of the affected areas.

In this model plan, the guidelines are:

- If it is expected that short-term protective action guides will be exceeded within two days of the start of a release, immediate evacuation is necessary.
- If projected daily exposure rates will be greater than one-tenth the short-term protective action guides, relocation must occur within two days from the time of measurement.

- If exposures that accumulate over longer periods of time are projected to exceed long-term protective action guides, then relocation is mandatory.

Relocation must be completed before short-term protective action guides are exceeded or within six months. The available time may be used to assess the feasibility of decontamination and to plan for relocation.

The decision to return The decision to return will be made by the Governor using the same long-term PAGs. Land and buildings that have radiation levels exceeding the long-term protective action guides may be used only for special purposes and are subject to controls limiting entry and duration of occupancy.

---

26. What arrangements will be made for reentry after an evacuation?

The organization of reentry depends on whether or not there has been a release of radioactivity. If there has been no release of radioactivity,

- the Governor will authorize reentry upon determinating that no further release will occur; and,
- given that decision, no restriction to reentry will be imposed;
- mobilization of state and local police will be adequate facilitate traffic flow and protect property.

If a release has occurred,

- All decisions authorizing reentry will follow a determination that there is no danger of further releases.
- Reentry into uncontaminated localities will be permitted after measurements confirm that no radioactive material was deposited.
- Reentry into areas in which the level of contamination is below long-term protective action guides will be permitted after sufficient measurements have been made to define the areas in which the protective action guides are exceeded.

- State and local police will facilitate traffic movement to re-opened areas and will maintain road blocks to isolate areas that remain closed.
- After the reopening of all areas where exposure does not exceed long-term protective action guides, state and local authorities will permit residents of contaminated areas to retrieve personal property under the following conditions:
  a. Appropriate state officials will accompany and guide persons temporarily entering the closed areas.
  b. Such reentry will be permitted only for time periods short enough to be safe according to short-term protective action guides.
  c. Any personal property retrieved will be monitored and decontaminated sufficiently to keep exposures below long-term protective action guides.
- The feasibility of decontaminating each area in which exposure exceeds the long-term protective action guides will be evaluated and recommendations will be made available to property owners and civil authorities.

MAINTAINING EFFECTIVE PREPAREDNESS

---

27. What public-education programs will be provided?

Public education is distinct from the system of public alert and notification. Public alert and notification involves relaying information to the public in the event of an emergency. Public education, by contrast, involves an ongoing effort, prior to any accident, to increase public understanding of and preparedness for a nuclear emergency.

In this plan, the educational program involves a broad mix of activities:

- in the *Inner Planning Zone*, members of the public will receive information distributed through the schools, public-information calendars, telephone-directory inserts, and brochures. They will also receive a biennial newsletter and have the opportunity to participate in special adult educational programs;
- in the *Middle Planning Zone*, the public will receive information through public information calendars, telephone directory inserts, and brochures; and,
- in the *Outer Planning Zone*, members of the public will receive a flyer giving basic emergency response information.

Goals. The primary goal of a public-education program is to enhance the effectiveness of public response during an emergency. Although any education program is limited in its ability to realize this goal, it may nonetheless contribute to a better informed and prepared public. The educational program should strive to realize several objectives:

- to increase public understanding about a host of issues (such as the nature of radiation, types of accidents, and possible adverse effects) in a balanced, accurate, and comprehensive fashion; and
- to familiarize the public with emergency-planning and -preparedness arrangements, including the emergency information system and the likely range of protective actions.

Roles and functions. The state emergency-management agency (PEMA in Pennsylvania) will be the lead agency responsible for developing the pre-emergency education program and coordinating the activities of the utility, the state, and local governments. The utility will be responsible for underwriting the cost of preparing and distributing the educational materials. Local citizens will play a central role in the development and evaluation of these materials.

The educational program will involve a broad mix of activities to ensure that the necessary information reaches as many members of the public as possible, including people who do not speak English.

Structure and organization The educational effort will be most intensive in the Inner Planning Zone. Here annual emergency public information calendars will be supplemented by inserts in the telephone directories and a biennial newsletter to all residents. Additional reminders about emergency-planning arrangements will be sent out periodically with utility bills. A noteworthy aspect of the program will include specially designed educational programs for teachers and school children. All real-estate transactions within the Inner Planning Zone will indicate the location of the property within this zone.

In the Middle Planning Zone, most of these information programs will also be offered. But intensive educational programs for school children and for adult education will not be included.

Beyond the Middle Planning Zone members of the public will receive only a flyer giving basic emergency response information, such as where to obtain further information. The public-information brochure issued to Inner and Middle Planning Zone residents will be available on request free of charge. Emergency workers in the Outer Planning Zone will participate in an educational program as part of their training (see Section 28).

Additional educational efforts will be tailored to the specific needs of special populations such as farmers, the Amish (in Pennsylvania), tourists and other visitors in the region, and institutionalized populations. Supplemental brochures will be issued to farmers, including many of the Amish. Institutions, such as hospitals, nursing homes, and prisons, will prepare, in cooperation with the state and the utility, information explaining institutional emergency plans. Tourists and other visitors to the region will have information in their telephone directory inserts as well as signs and posters in parks, campgrounds, and other public places. Their attention will be drawn to those materials by posters on the back of hotel doors, and in gas stations, phone booths, and other prominent public places. Road signs will also indicate the boundaries of the emergency-planning zones.

Widespread public participation in developing and implementing these informational and educational programs will be extremely important for community acceptance of the planning arrangements. The public will be involved through a Local Review Committee and public hearings (see section 28).

---

28. What training and evaluation programs will be required to maintain a high state of preparedness?

A difficult problem for emergency planning is maintaining a high state of preparedness in the absence of accident events. Routine can easily erode the effectiveness of a planning effort. To overcome this problem and to maintain a high state of readiness, this plan calls for:

- training programs for all emergency workers, including volunteers;
- the establishment of a Local Review Committee to evaluate the plan and its training programs; and,
- a system of drills and exercises to increase interorganizational coordination, train emergency workers, and test the effectiveness of the plan.

Training programs. Training programs have several objectives:

- to provide comprehensive training for *all* emergency workers, including volunteers;
- to train each worker to understand his or her individual role and responsibilities and how they fit into the overall emergency response system; and,
- to encourage flexibility and adaptability by ensuring that workers know *why* they and others are designated certain responsibilities and *what* essential functions must be performed.

An introductory training program will be developed for all full- and part-time professional and volunteer emergency workers. This program should also be made available to all institutional staff with emergency-response roles in hospitals, prisons, nursing homes, schools, and colleges. The training program will provide a basic introduction to the nature of nuclear accidents, types of radiation, health effects, protective actions, monitoring and decontamination, reentry and relocation, communications, and overall emergency planning and response. The use of detecting equipment will be part of the training. This training will be most intensive in the Inner Planning Zone but will also be used in the Middle Planning Zone.

More specialized programs, tailored to the specific responsibilities of particular groups of emergency workers, will supplement this introductory training program. Thus, medical personnel will receive detailed training on: radiation health effects and their treatment; screening, monitoring, and decontamination procedures; and medical triage (see Sections 20 and 21). Similarly, school teachers will receive specific training in their roles and responsibilities during nuclear emergencies, and especially the need for and methods of administering potassium iodide (see Section 22). In particular, these training materials will emphasize not only what actions should be taken but also *why* such actions are necessary.

State officials will develop the introductory and specialized training programs, with substantial participation of local communities and appropriate representatives of emergency workers. The utility will provide support for the training programs.

Drills and exercises. These constitute the most important mechanisms for evaluating emergency planning and preparedness. Drills are designed to test on-site emergency preparedness and procedures. Exercises are designed to test off-site emergency-response capabilities. The schedule, scope, and content of exercises will undergo evaluation by the Local Review Committee as well as the state emergency-management agency, the utility, and various local agencies.

In this plan, exercises will:

- take place at least once in the first year after the plan is approved, annually in the Inner Planning Zone, and at least once in each subsequent two-year period in the Middle Planning Zone.
- be sufficiently comprehensive to include all categories of participants (i.e., emergency personnel at all levels).
- be conducted at varying seasons and times of day.
- be conducted without prior announcement once every two years in the Inner Planning Zone and once every six years in the Middle Planning Zone.
- be evaluated by the Local Review Committee, which will disseminate its evaluation findings to the local media and make recommendations for improvements to state officials.

Evaluation. The Local Review Committee, appointed by and reporting to the Governor, will provide continuing evaluation of the plan and the state of emergency preparedness. This committee, composed of 15 members of the public, will be independent of the state emergency-management agency, will have its own budget (from funds provided from the utility), and will be authorized to hire its own consultants and a small technical-support staff. The committee will represent local citizen concerns about emergency planning and preparedness and will give particular attention to

- the needs of special populations;
- the warning and communications systems;
- the operation of screening stations and decontamination and mass-care centers;
- reentry and relocation procedures; and,
- any other matters of local concern.

The primary responsibilities of the committee will be to

- evaluate the plan against the stated objectives and regulatory criteria.
- evaluate the schedule, scope, and content of exercises.
- review performance on drills and exercises.
- participate in the development of training programs and assess their effectiveness.
- make formal findings concerning the adequacy of overall emergency planning, implementation, training, and resource allocation, in the light of formal performance criteria.

---

29. How will the state of emergency preparedness affect plant operations?

In recognition of the role of emergency planning as the "fifth layer of safety" for a nuclear plant (see Section 7), the nuclear plant's continued operation will depend upon the state of emergency preparedness as follows:

- If the state of emergency preparedness becomes temporarily seriously degraded, the plant will be shut down until normal conditions have been restored; and
- continuance of the plant's operating license will be subject to a five-year review of relevant emergency response capabilities.

Emergency preparedness as a plant safety system. Emergency preparedness is not an afterthought, but the final layer of safety for a nuclear plant. Accordingly, emergency preparedness must meet the same standards as safety systems in the plant itself. This model plan meets that requirement by modification of earlier regulatory practices, so that either temporary serious degradation or prolonged breakdown of emergency preparedness will command appropriate restrictions on plant operation.

Temporary degradation of emergency preparedness. A variety of events may lead to significant short-term degradation of emergency-response capabilities. For example, the computer or communication systems that provide plant data to state authorities may be temporarily inoperative. Similarly, a severe snowfall, flood, or hurricane may seriously impede evacuation of surrounding populations. Under this plan, such contingencies are anticipated and written into the plant's Technical Specifications as conditions that may require a power reduction or plant shut-down. The precise nature of those conditions will be determined as part of the US Nuclear Regulatory Commission's licensing process.

Emergency preparedness and continuance of the plant's operating license. To guard against slow deterioration in emergency-response capabilities, the continuance of the plant's operating license will be subject to a satisfactory five-year review of the status of emergency preparedness. Under this plan, the US Nuclear Regulatory Commission's licensing criteria would be changed to include this requirement. The Commission will be required to take account of formal review findings by the Local Review Committee and by the Federal Emergency Management Agency.